AF477166

# Environment and We in 21$^{st}$ Century

# The Author

 **Dr Poonam Agarwal** has obtained her M.Sc. and Ph.D. degree in Botany and got selected for post of Assistant Professor, Botany through Uttar Pradesh Public Service Commission, U.P. She is currently working as Associate professor and Head, Department of Botany, Dr. Bhim Rao Ambedkar Government P.G.College, Fatehpur (U.P.). She has been teaching basic and advanced Botany for last twenty three years. She has to her credit 35 research papers published in National and International Journals, books and magazines. She has frequently participated and presented papers in several national and international seminars. She has written a book entitled 'Challenges to Biodiversity Conservation and Sustainable Development' published by Scholars World, Astral International (P) Ltd., New Delhi. The book entitled 'Environmental Crisis and Society: A Sustainable Approach' and special issue of the journal 'International Research and Reviews' are also edited by her as one of the editors. She has the life membership of Indian Association of Music Therapy working as Executive Body Adviser of the Association. She is also assigned as Associate Editor of World Journal of Applied Sciences and Research and Editor of International Journal of Music Therapy.

# Environment and We in 21$^{st}$ Century

Dr Poonam Agarwal

2019

**Scholars World**

*A Division of*

**Astral International Pvt. Ltd.**

New Delhi – 110 002

*Published by* : **Scholars World**
*A Division of*
**Astral International Pvt. Ltd.**
– ISO 9001:2015 Certified Company –
4736/23, Ansari Road, Darya Ganj
New Delhi-110 002
Ph. 011-4354 9197, 2327 8134
E-mail: info@astralint.com
Website: www.astralint.com

*Digitally Printed at* : **Replika Press Pvt. Ltd.**

*Dedicated to*

*God*

*Who always guided me..........*

# Preface

Environment belongs to all and is very important for all of us. The condition of our environment is depleting regularly. As the most intelligent being on earth, man has constantly tampered with nature making use of his inventive genius and disturbed the eco-balance sowing the seeds of his own peril. It is only the closing decades of 20th century, environmental issues have emerged as a major concern for the survival and welfare of mankind throughout the world. Modern civilization, armed with advancing technology and fast growing economic system is under threat from its own activities leading to pollution, global warming, depletion of natural resources, biodiversity loss etc. The fast eroding life support system of our biosphere not only endangers the future of our next generation but it is a major threat to our life itself. The root cause of all these environmental problems is human population explosion. To fulfill the basic needs of people, unplanned industrialization, colonization, modernization, motorization, mechanization and modern agriculture on one hand and increasing pressure on life support system (air, land and water) on other hand have created increasing pressure on natural resources in the present century. Disastrous condition of environment is alarming the whole world to take it seriously. We are accelerating to the point of no return. Nature in itself is non-renewable, so is the man's life.

The anthropological activities have led the situation to a crisis stage and technological solutions of crisis always create new and more serious problems. Time has reached when we are facing challenge to our intellect and wisdom for survival of humanity. We, Indians have a long tradition of reverence, respect and love for nature and we must continue this tradition. The nature has transformed into culture, therefore, there is an interactive relation between nature and culture. Sustainability characterises the management of nature achieving ever lasting satisfaction of human needs, improving the quality of life and refusing to discriminate between the means and the end. There must be rational use of resources including environmental consequences, both in long and short terms. Now, it is dire need of the day to awake, to respect and to pay special attention to our beloved mother earth deserves.

The sustainable development can not be achieved by any government at its own level until unless we people have a participating role in it. Therefore, every individual should be aware about the fact that degrading our environment means actually harming our own selves. Awareness for a safe and cleaner environment should be part of formal or informal education. We must have to change our life style in such a way so that contribution towards environmental pollution is minimum making judicious utilization of natural resources. Environmental laws can be applied for improved environmental performance. If we fail to do so, the coming generations will blame us for robbing them of their life support systems.

The various chapters presented in this book reflect all different aspects of environmental abuse and degradation along with environmental preservation and protection. All these topics have a concern for all living beings including us also. This led me to believe and sure that this book will be both timely and relevant for students, researchers, young scholars and people of general interest to find an answer to a number of important environmental related issues in 21$^{st}$ century.

I would like to express my sincere thanks to Almighty God, the source of all knowledge, understanding and wisdom. From Him, I owe all that I have and all that I am.

Every challenging task needs self efforts as well as guidance of elders and younger especially those who are very close to my heart. So, I take this opportunity to place a deep sense of gratitude to my family members who have been my constant source of inspiration during preparation of this book. Our deep indebtedness and gratitude are due to all those people who have been cooperative and supportive in different ways in bringing out this book. I would also like to thank to all those, living or non-living, who have been directly or indirectly related to my work. Finally, thanks are also due to Scholars World a unit of Astral International (P) Ltd, New Delhi for publishing this volume.

*Dr Poonam Agarwal*

# Contents

# Chapter-1

# Our Environment

Environment is everything that is around us. It is a complete set of physical, chemical, ecological, geographical and natural forces that surrounds an individual or organism. All living things live in this environment naturally. They constantly interact with it and adapt themselves to conditions which ultimately determine form and nature of their survival. An environment, therefore, is the natural surrounding which help life to grow, nourish and destroy on this earth. The earth is the home of we people, hence the soil upon which we live and the surrounding life supporting environment is of vital importance to us. The word 'environment' is derived from a French word *'environner'* which has the basic meaning 'that which surrounds'. The biologist Jacob Van Uerkal (1864-1944) introduced the term 'environment' in ecology. Everything comes under an environment, the air we breathe, the water we use for daily needs and drink, plants, animals, other living things, and sunlight etc. around us. An environment is a complex of many factors which interact not only with the organism but among themselves. Thus, environment is the product of biophysical, geographical and social structures and processes which we are a part of.

The earth along with the environment, which supports life of living organism is called biosphere. Biosphere extends to about 7 km. into the earth surface, into ocean depths up to 10.67 km. and upwards into the air to about 10 km. However, the entire biosphere is not habitable and there are certain regions where life can not exist normally. The region of biosphere which is not favorable for life, such as volcanoes, Polar Regions, extreme deserts, hot springs etc, is termed as 'Para biosphere'. The regions suitable for normal life is called as ' Eubiosphere' which covers 'Atmosphere'- gaseous in nature and extends around the earth; 'Lithosphere'- solid crust of rock and soil on earth's surface; 'Hydrosphere'- the depressions in earth crust filled with water. The environment we live in represents the interplay of various segments of biosphere. The interaction is promoted by energy input supplied by sun.

In nature, there is a continuous circulation of water from earth to atmosphere and from atmosphere back to earth. It is controlled by water / hydrological cycle. Rocks that exposed on surface of earth are eroded by wind and water and deposited in ocean. Thus, masses of rock materials are transformed from continents to oceans and back to continent in a cyclic fashion. It is Rock cycle. Biochemical cycle links the rock cycle with that of water cycle and refers to transport and transfer of dissolved and solid substances through our environment on a time scale representing natural reactions with involvement of rock or soil, water, atmosphere and to some extent the biosphere. An environment is healthy environment when all natural cycles go side by side without any disturbance. Any disturbance in nature's balance affects environment totally which ruins the human beings too.

A clean environment is very necessary for us to live a peaceful and healthy life. But our environment is getting disturbed as well as dirty day by day due to some negligence of we people. Modern civilization, armed with rapidly advancing technology and fast growing economic system is under increasing threat from its own activities. Now, in the era of development and technological advancement along with advanced standard of living of people, our environment is getting affected to a great extent by means of air, water, soil, noise and chemical pollutions, deforestation, agricultural developments and other disasters resulted by some bad and selfish activities of human being. The fast eroding life support system of our biosphere not only endangers the future of we people but it is creating the battle for survival of life itself.

## Environment – Today and Tomorrow

For millions of years, man and nature have coexisted in harmony. There were few environmental issues in early agricultural society. As man learned to use his intelligence to bring about change through what we called as technology, he started to surpass nature. Later on in 18ᵗʰ century, this technological advancement is named as 'Industrial Revolution'. As the most intelligent being on earth, man has constantly tampered with nature and disturbed the ecological balance. The dominance of machines over humanity and nature has altered the environment and landscapes on a scale. In recent decades, many environmental problems have increased due to human activities and unplanned management of technological developments those interfere ecosystem. Various human activities have induced many undesirable effects to the environment which may be a threat to economy, natural resources, human health, gene pool of ecosystems and air, water, soil, and various types of pollutions etc. For the last 20ᵗʰ century, average temperature of planet earth was increased by 0.6 °C and it is still rising. According to fourth assessment report of Intergovernmental Panel on Climate Change (IPCC), the projected temperature will be 1.8° C to 4.0°C by the end of 21ˢᵗ century. The root cause of global warming is green house gases. Naturally occurring green house gases in the atmosphere like carbon dioxide, methane, water vapours and ozone trap some of the heat and maintain a mean warming effect of about $33^0$C without which the earth would be inhabitable. Therefore, due to anthropological activity only green house effect is intensified leading to global warming along with 3/4 of increase in carbon dioxide from fossil fuel burning. Rest is due to land use change especially deforestation and urbanization. Forest are known as 'Carbon Sinks' because trees act as absorber of carbon dioxide in the atmosphere and release oxygen in return. The unplanned deforestation have reduced the concentration of oxygen in the atmosphere and raised the temperature of the earth. All these things are happening today as a result of man's past and present experimentation on nature.

The global warming results in adverse environmental conditions that threaten the biodiversity, ecosystem balance and also socio- economic life of humans. Most remarkable impact of global warming is climate change which in turn could lead to extreme weather conditions like temperature drop and rise. The warmer places may become cool or may be warm places might be even warmer. The similar situation could happen to cold places also. Rise in global temperature will significantly caused melting of ice at Arctic. It is predicted that by 2050, entire Antarctic region will be totally ice free. According to United Nation Climate report, Himalayan glaciers could disappear by year 2035. As temperature rises, the Ganges and Indus rivers, the largest rivers of Asia, which depend on drainage from Himalayas could get less water. Therefore, billions of people from surrounding countries would have to face water shortage. On the other hand, drought, heat waves, wildfires would increase as compared today. Due to changes in weather conditions, Cherrapunji in Meghalaya, India having the highest rainfall is now facing desertification. Due to melting of ice in Polar Regions, rise in sea level is another effect of global warming. Because of this, some land areas or cities situated near sea coast might become submerged. Studies report that 18-35% of planet's flora and fauna will get extinct by the year 2050. Animals are likely to move to areas near Polar Regions due to warming in tropical and subtropical regions. Insufficient rainfall and longer droughts are expected to affect crop yield. Even the increase in the form of natural calamities resulting out of climate change are indirectly brought by human beings. Warm climate and flood may intensify the risk of spread of pests and vector borne diseases. High temperature in rivers will encourage growth of algal bloom causing increased Biological Oxygen Demand (BOD).

While developing technology for benefit of man, it is also mandatory not to overlook simultaneous impact on environment because one day, we should pass it on to our future generations as their legacy. Now, environmental problems have become not only a public health issue but a global crisis that may threaten the future of our living planet earth. We are witnessing a real crisis due to our wrong choice.

We human consider ourselves the most intelligent species on this planet earth......but are we? Animals, not as intelligent as we are, are not doing any activity which is deleterious to the environment. But we are doing it brutally. We do have all the chances to choose their acts but not its every consequence. Nature is calling out to us but we are ignoring the warning. Global warming is like diseases made by man to kill gradually not only the earth but himself also. We are reaching the point of no return. Nature in itself is non-renewable, so is man's life. This alarming situation needs to be overcome by contribution to a better understanding of people and environment and to a more balanced relationship between the two. It can be done only through reorienting and modifying our attitudes towards natural environment in which we live. The environment can live without us but we can not live without environment. The presence of nature is ideal to man's self, therefore, need to get closer in that kind of ideal partnership. We should treat our environment in a more relevant way.

## Chapter-2

# Population and Environment

Global human population is increasing at an alarming rate. Within next 50 years, the developing countries will be double or triple in size if they cannot stabilize their population. Most developed countries currently consume natural resources much faster than they can regenerate. Most developing countries with rapid population growth face need to improve living standards. We exploit nature to meet present needs, are we destroying natural resources necessary for the future? Many feel that wars to be fought in future will be for natural resources. In future, if present pattern continues, as resources get depleted and wasted in wars, additional conflicts will arise for the limited resources. Now question is – does human population affect and put stress on the environment and society?

A sustainable India is one that allows its people to live socially engaged and prosperous lives in a healthy environment. Population plays a decisive role in safeguarding the environment and sustainable development. The burgeoning population of India is imposing an increasing burden on its limited and continuously depleting natural resource base. The huge increase in human populations has resulted in a substantial degradation of environmental conditions. The human populations can exert strains upon their natural surroundings is nothing new. The industrialization, agricultural improvement and energy policies have improved the living standards during last four or five decades. But now these are starting profound environmental impacts that can adversely affect the natural planetary cycle upon which all life depends. Not only the number of people but also the life style, consumption pattern and region people inhabit and use directly affect the environment. The relationship between population growth and environmental degradation may appear to be rather straight forward. More people demand more resources and generate more waste. One of the challenges of a growing population is that the mere presence of so many people sharing a limited number of resources strains the environment. Population has contributed to some of the environmental problems that we are experiencing today-

☆ Depletion of ozone layer- protects humans, plants and animals from UV radiations.

☆ Global climate change- rise in global temperature could raise sea levels, may bring flood, drought, altered rain fall patterns etc.

☆ Loss of biodiversity- many plant and animal species are under threat of extinction.

☆ Deforestation- forests are vital to maintain healthy ecosystem.

☆ Desertification and soil erosion.

☆ Pollution of air and water.

☆ Worldwide diffusion of hazardous substances including persistent organic pollutants that adversely affect human, animals and plants.

☆ Fresh water crisis- demand of fresh water is soaring as population grows and use per capita rises.

☆ Public health- unclean water, poor sanitation, air pollution, heavy metal contaminants cause health problems.

## Population Growth and its Impact on Environment

Rapid population growth and the consequent changes in demographic structure and uneven population distribution are the factors that impose pressure and stress on economic development, the environment and natural resources as well as social conditions. Population explosion is seriously threatening the delicate balance between human being and their environment. Deforestation, desertification and water scarcity are already having devastating effects. The complex influence of population dynamics on environment and their interrelationship are emphasized mainly in three specific areas- Land use pattern (Forest), Water resource management (Fresh water) and Global climate change.

## Forest

Forests –"The Earth's Lungs" have been the base of development of human civilization. They have multifaceted utility with renewability through tree plantations. Forests products like wood products for housing, furnishing, fuel, packaging; food products and medicinal herbs play major role in global economy. Forests have many functions of value both to humanity and to the nature itself. They absorb carbon dioxide and provide oxygen, anchor soils, regulate water cycle, protect against erosion and provide a habitat for millions of people, animals and plants. The social, economical, recreational, aesthetic and spiritual aspects of forests can not be ignored. And as the store house of carbon, forests are keys to regulate climate. They also intercept sound, thus reduce noise decibels. In spite of their endless uses, why forests are destroyed at all? It is well illustrated in following (Figure 1.1.).

75% of loss in global forested area has occurred in 20ᵗʰ century but since 1950, the advent of mass consumption of forest products have quickened the rate of deforestation. It leads to huge release of carbon. Besides, growing resource requirement requires some form of land use change – the expansion of food production through forest cleaning to intensify production or to develop the infrastructure necessary to support increasing human number. More people need more food, which can come only from either expansion of agriculture into new lands, or use of existing agriculture land more intensively or a combination of two depending on relative costs, themselves a function of technological, institutional and ecological factors. These types of land use changes have ecological impacts. Land conversion to agricultural use can lead to soil erosion and the chemicals often used in fertilizers and pesticides can also degrade soil. Deforestation can lessen the ability of soil to hold water, thereby increasing the frequency and severity of floods. Human induced changes in land use often results in habitat loss, the primary cause of species decline. A World Bank study found that rate of soil loss was ten times higher on forest lands where slash and burn shifting cultivation was practiced than in undisturbed forests. If, to accommodate and to meet the demands of population, the current rates of forests cleaning will continue, one quarter of all species on earth could be lost within next 50 years.

As population grows and per capita consumption of forest products increases, countries must do more to manage forest resources on a sustainable basis. Many organizations urge Government with large forest resources to enforce legislation and to introduce more effective forest conservation initiatives. Millions of people rely on forest products for their livelihood. Sustainable forest management will require not only enforcement of laws but also alternative sources of livelihood for many rural people.

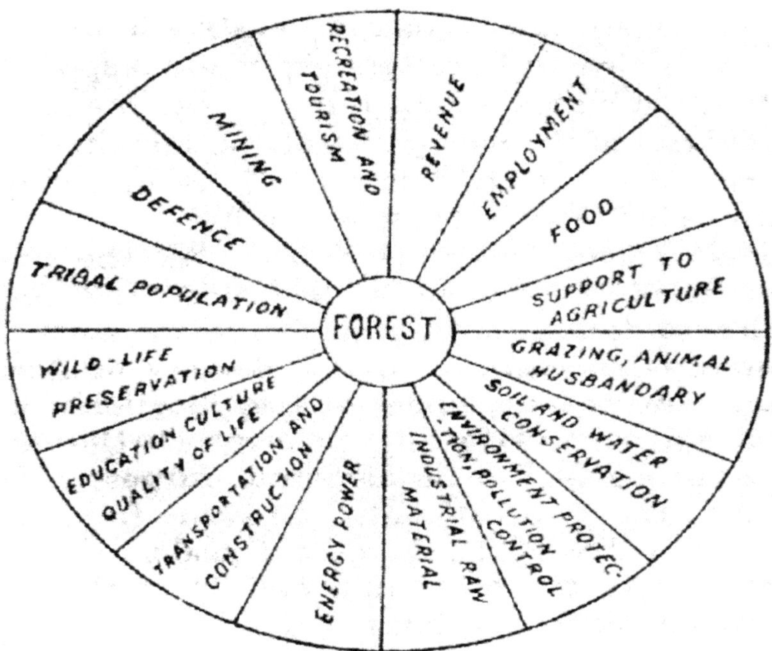


**Fig.1.1: Various uses of forest**

## Fresh Water Management

Most characteristic element of our planet is undoubtedly water. Indeed, more than 2/3 of earth's surface and 70% of human body has water- indispensible for all living organisms. Life is impossible without water. As population continues to grow rapidly and use per person rises, demand for fresh water soars. Yet there is no more water on earth now than there was two thousand years ego when the population was less than 3% of its current size. Beyond the impact of population growth itself, the demand for fresh water has been rising in response to industrial development, increased reliance on irrigated agriculture, massive urbanization and rising living standards, domestic consumption etc.

Supply of fresh water available to human beings is shrinking because many fresh water resources have become increasingly polluted. Chemical run off from fertilizers and pesticides, industrial effluents also damage water resources. Pollution originating from human waste especially located too close to water supply sources affects both surface water as well as ground water. Polluted water causes various diseases as malaria, cholera, typhoid etc. Over use and polluted water supplies pose increasing risk for many species of flora and fauna.

Whether water is used for agriculture, industry or municipalities, there is much room for conservation and better management. The world needs a blue revolution to conserve and manage fresh water supplies in the face of growing demand from population growth, irrigated agriculture, industries and cities. A water short world is an inherently unstable world. Managing our water resources with care and intelligence for the use of present and future generations is a major responsibility which has to be shared by Government and public alike. Competition for fresh water supplies breeds social and political tensions. At the dawn of 21^{st} century, there is a growing risk that wars will be fought for fresh water. It is time for wide spread conservation and rain water harvesting measures with effective water management policies.

## Global Climate Change

Over the last ½ century, carbon emission from fossil fuel burning expanded at nearly twice the rate of population boosting atmospheric concentration of carbon dioxide and green house gases. These induce changes in global climate, sea level rise and coastal wet land ecosystem destruction. Recent years have among the warmest on record. Research suggests that temperature have been influenced by growing concentrations of CFC gases which absorb solar radiations and warm the atmosphere. The destabilization of our climate threatens more intensive heat waves, more severe drought and floods, more destructive storms and more extensive forest fires. Related shifts in rain fall and temperature may affect food production, biological diversity, ecosystem as well as human health. The carbon contribution from forest will likely increases in coming years as the burgeoning human population continues to cut down the forests.

Since human habitat and livelihood are closely interconnected with surrounding ecosystem, increased pressure from population can cause irreversible damage to fragile ecosystem. Significantly in some cases, technological advances of developed countries lead to more exploitation of natural resources. There is growing imbalance between the demands of human population and resources that support human life.

It is a fact that we are using and sometimes abusing most of the resources we have or better to say we had. Deforestation for living space, wildlife poaching for human benefit, industrialization to support human needs etc. are just a few of the habits that we have adapted in the recent past without realizing how it would affect the earth and us in long run. Can enough food and energy be provided- not to mention job, education, health care and waste disposal to accommodate these billions without causing some irreversible ecological collapse.

The large increase in population has also had several negative social effects that interact with technology and the environment. Because most sources are consumable and non- renewable, the increase in population means resources are consumed and depleted more rapidly. Such high consumption of food, shelter, material goods is associated with increased level of pollution, for example- as more people drive cars demand for more oil as well as air pollution level increases. The rapid increase in human population combines with desperate poverty and rising levels of consumption are depleting natural resources on which the livelihood of present and future generations depends.

Population is recognized as an indirect driver of biodiversity loss, as human demands for resources like food and fuel play a key role in driving biodiversity degradation. This happens primarily through the conversion of ecosystem to food production. Household demographic factors, such as household size, have important implications for resource consumption, with rapid increase in household numbers associated with loss in biodiversity. Over exploitation and habitat loss as a result of population and other pressures is likely to contribute to a high risk of extinction of plants and animals. Agricultural land expansion is the most dominant driver for habitat loss, which combined with unsustainable forest management, contributes to the greatest cause of species moving closer towards extinction. Increased demand for goods and services to meets the needs of a growing population will undoubtedly exert more pressure on the components of biodiversity – ecosystems, genes and species.

The pressure on environment intensifies everyday as the population grows. As the 21$^{st}$ century begins, growing number of people and rising level of consumption per capita are depleting natural resources and degrading the environment. Though the relationship is complex one, population size and growth tend to accelerate the human impacts on the-environment. If the current expansion continues in future, it will cause overall scarcity of resources.

There is urgent need to bring balance in population growth, socio-economic development, natural resource consumption and environmental sustainability. Despite the wide spread negative effects of human population increase such as industrial growth, resource depletion, ecosystem destruction, global warming, poverty , humans have met with many success in solving or lessening some of these problems. "Green Chemistry" practices have sought to reduce chemical pollution and provide cleaner methods of industrial production. Government policy regarding the use of Chlorofluorocarbons' has significantly reduced the size of ozone hole.

Technological advances and education programmed on birth control have begun to slow the birth rates and stem overall population growth. Researches must continue to search for new environment safe methods of sustaining the burgeoning human population before these problems reach the threshold of catastrophe. Slowing population growth will not only ease off pressure on biodiversity, but will also empower women and their families. Living in harmony with nature has been an important feature of cultural heritage of India and this healthy tradition should not be ruptured.

# Chapter-3

# Urbanization Stress on Environment

Urbanization is the physical growth of urban areas as a result of global change. It refers to general increase in population and the amount of industrialization of a settlement. It includes increase in the number and extent of cities. It is one of the most significant trends in the modern world. Urbanization is closely linked to modernization, industrialization and the sociological process of rationalization. It has also changed the traditional pattern of social life and nature of human communities. The towns and cities in developing countries are growing faster than ever before. It is increasing in both the developed and developing countries. Some of the so-called mega cities have more than twenty million inhabitants. The global proportion of urban population rose dramatically from 13 % (220 million) in 1900, to 29% (732 million) in 1950, to 49% (3.2 billion) in 2005. According to the UN World Urbanization Prospects Report, the figure is likely to increase to 60% (4.9 billion) by 2030. However, more than 90% of future population growth will be concentrated in cities in developing countries. The urbanization has become a major demographic issue in the 21$^{st}$ century not only in India but all over the world. Although the larger share of the population increase in the cities of developing countries is caused by the children of people already living there, the rural – urban migration continues unabated. Nowadays, nearly half population of the globe is residing in cities, urban or periurban areas. Increase in sources of livelihood, better job conditions, modern infrastructure facilities, better professional and technical higher education, better health facilities and many other avenues have made cities attractive for one's future prospects. Ultimately, people have started migrating from rural to urban cities. It is estimated that in the next half century, about 70% of population would be urbanites. And most of the urban population growth will take place in developing countries, led by Asia. In India, the urban population has increased nearly five times due to migration. The pace of urbanization will be fast if the industrial growth is fast or in other words, it is not possible to have industrialization without urbanization. This increasing degree of urbanization has emerged as a growing burden on cities as well as on environment. People's concentration in urban areas resulted both positive and negative impacts. On one side, it resulted in educated and cultured society, more job opportunities, intensive agriculture, industrialization and mass production of goods improving the economic condition of people at regional, national and global level. On the other side, as population concentrates in cities, the demand for natural resources, land, water and electricity increases manifold. Side by side, cities also have more carbon emission, more raw materials for industrial use and more domestic water use etc. Urbanization has disturbed the ecological balance by creating air, water and noise pollutions through industrialization. Haphazard growth of urban centers and their ill- planned urbanization aggravated the problems of pollution and deterioration of the environment. It has been revealed by a survey of major cities of India, there has been substantial increase of the suspended particulate matter (SPM) in the air, suggesting the presence of dust and carbon particles coated with toxic gases. These high levels of air pollution are largely attributed to incomplete combustion of diesel and leaded petrol by the vehicles. Therefore, urbanization process in India has proven to be more pollution-intensive than that in the old industrial nations of Europe and North America.

As cities expand to accommodate the increasing migrants, forests are converted into concrete jungles. The forests remain as small pockets of trees and vegetation (urban forests) in and around residential human societies leading to disastrous ecological crisis in our towns and cities. Therefore, housing society planners have started developing green cover of 'urban forests' wherever land is available to cope up the sociological and ecological needs of city dwellers. Global cities like Beijing, Malaysia, New York, Paris and Singapore etc. follow urban forest planning of availability of 9 meter square green space per city dweller as estimated by WHO and FAO. However, there is no authentic data for green cover area in urban India. In 2009, a survey conducted by Forest Survey of India (FSI) has opined that Chandigarh tops in having highest geographical green area followed by Bangalore, Pune, Bhopal and Delhi. Therefore, a challenge for biodiversity conservation is imposed due to urbanization. Developing societies and townships come into existence at the cost of agriculture while agriculture develops at the cost of green trees. The conversion of farm land  and watersheds for residential purposes have negative impacts on food security, water supply as well as the health of the people., both in cities and in peri-urban areas. Besides, loss of forest cover robs wild animals of their homes bringing them into conflict with humans. Besides, the cut of a single tree causes the loss of habitat of many species of birds, insects, micro flora and micro fauna. The cities are witnessing decline in bird population due to loss of green cover. Scientists have discovered that Indian palm squirrels (*Funambulus palmarum*) are using plastic threads and plastic bags in urban areas as nesting materials instead of leaves, twigs, shredded barks, mosses and other soft materials. The proportion of anthropogenic (artificial) material used in nest building is directly related to the extent of urbanization. Although Indian palm squirrel usually built nests using natural materials, they appear to be adapting themselves to changes in habitat by using most available plastic material to survive. The research study from Bio-Psychology Laboratory and Institute of Excellence, University of Mysore, explains that use of plastic materials for nest building by palm squirrel is an example of the struggle for existence in altered habitat. On the other hand, increase in agricultural chemicals, industries and transportation have resulted an increase in pollution of land, air and water. Food chain and food web have affected due to DDT like hazardous chemicals and fertilizers. Flowering time and its pattern has also affected due to increase in temperature. The changing climate is disturbing the relationship between plants, pollinators and symbionts. In metropolitan cities, deforestation has resulted in increase of particulate matter in the air, thereby enhancing the health problems in urban dwellers.

Plants reduce the air temperature by providing shade reflecting the sun light and also remove air pollutants efficiently. Noise from factories, automobiles, house appliances etc. is another major problem in urban societies. Studies have confirmed that leaves and branches of trees intercept noise and absorb noise energy. Most important ecological phenomenon rendered by the trees is the maintenance of hydrological cycle. Besides, the roots of trees act as filters and reduce the contamination of ground water. Therefore, dire need for most of the cities in India is to bring more and more available area for plantation and conserve the existing biodiversity.

The bulk of fish production of Bengal comes from sewage –fed fish ponds of East Kolkata wetlands. Not only fishes, here at the wetlands vegetables and paddy are also produced in considerable quantity. But today pressure from urbanization is the greatest threat to these unique wetlands and its aquaculture practices. Now, it is high time that the wetlands and their ecological, economical and human friendly model are protected from pressures of urbanization. Otherwise Bengal will be a loser and 'Bengal Food Bowl' of East Kolkata wetlands will vanish away completely.

The only way by which cities may able to meet the challenges present before urban modern dwellers is to adopt proper urban planning and to maintain a rich green tree cover. Tree cover has more ecological, sociological and environmental importance in densely populated urban cities than cleaner green environment of villages. Setting up of green cover with scientific inputs can also help in biodiversity conservation in urban cities. Legislation can also play a vital role in biodiversity conservation in urban areas. Implementation of urban tree act and declaration of protected area are another steps towards conserving diversity in modern cities. The most important step towards conservation is to create and increase awareness in all sections of the societies. Residential, commercial and industrial areas must be properly planned. Importance must be given for locating sufficient number of public parks and gardens.

What is important then is not to question the validity of city development, but to convert cities and towns a better place to dwell in. Good governance and more attention to environmental hazards caused by congestion are some of the demands that must be met to cope with the problems of cities. To handle this situation and to save planet earth and people's health, there emerged the need of well planned and managed cities, so called Smart cities' having high living standards with remarkably low levels of energy consumption, natural resource use and wastes. Therefore, idea of 'Smart city' emerges i.e. able to solve urban issues paying attention to the environment. Smart cities are now becoming a reality. Emerging new technologies are poised to reshape our urban environment utilizing wireless network, web technologies, mobile based information and communication technologies and ultra power sensors. So, NDA government has a proposal to build smart cities in India. To accommodate rapid urbanization, the Govt. of India has allocated Rs. 6000 crores for the smart cities project to develop infrastructure in cities.

Smart City is a developed urban area that creates sustainable economic development and high quality of life in various sectors such as economy, transport, environment, health care, people's living and government excelling in these areas through human capital, social capital and/or information and communication technology infrastructure.

'Smart City' is a multidimensional phenomenon and its model is based on six main dimensions of development( Figure 2.1) -

  (i)   *Economy:* Ability to create employment of innovative companies, good quality universities and advanced research institutes.

  (ii)  *Environment:* Intelligent use of resources, promotion of sustainable development based on recycling and waste reduction. ICT enabled energy grids, pollution control and monitoring, green urban planning.

        Urban services such as street lighting, waste management , drainage systems and water resource systems that are monitored to evaluate the system, reduce pollution and improve water quality.

  (iii) *Governance:* Adoption of policies for boosting development and inter-municipal networking capacity.

  (iv)  *Living:* Advanced services for improving the quality of life.

  (v)   *Mobility:* An innovative efficient system of public transport.

  (vi)  *People:* ICT enabled life style. Citizens of city are active and participate in public life.

## Environmental Sustainability

Sustainable development in 21ˢᵗ century will to a large degree depend upon how cities, towns and villages everywhere interact with the environment and utilize natural resources. Cities have to face a number of environmental sustainability challenges generated by the city itself or caused by weather or geological events. One of the first stages to address sustainability is to increase resource efficiency in all domains such as energy efficiency in building and networks, fuel efficiency in transport, water efficiency and new methods to transform waste to energy. Technology is not only aspect required for sustainability, but is an important and necessary step forward. The other pressure arises on quality of life and health. Not only on citizens and authorities are environment aware but the economic implications of pollution can be serious due to impact of health and attractiveness for business to operate from the city. Another pressure is for risk management and resilience to environmental shocks (such as heat waves and flooding caused by climate change). It needs to be integrated in city planning, based on estimated future risks. The integration of different technologies in areas of Information and Communication Technology (ICT), transport, energy, water etc. which form the infrastructure backbone of cities, currently offers the best prospects for sustainability. Therefore, vision of 21st century should be –"Ensure sustainable development while preserving the environment and achieve balance between economic and social development. Focus on improving the quality of air, preserving water resources, increasing the contribution of clean energy and implementing green growth plans". Various projects for achievement of a smart city or an environmental futuristic city are being advanced in Japan and all parts of world. ICT is expected to be an effective tool for monitoring, visualizing, analyzing and optimizing flows of resources, energy, information, persons and goods. Sustainable city Network may work as a new approach to solving various problems with ICT and achieve optimization by having cooperation in large areas in and between cities, towns and villages.

## Eco cities

It is a city built off the principles of living within the means of the environment. The ultimate goal of many eco cities is to eliminate all carbon waste to produce energy through renewable sources and to incorporate the environment into the city, however eco cities have intentions of stimulating economic growth, reducing poverty, organizing cities to a higher population densities and therefore, higher efficiency and improving health.

## Smart Sustainable Cities

"A Smart Sustainable City is an innovative city that uses Information and Communication Technology and other means to improve quality of life, efficiency of urban operation and services and competitiveness while ensuring that it meets the needs of present and future generations with respect to economic, social and environmental aspects". It can only be done only if we have our own dream of a modern Indian city. Emerging technologies are poised to reshape our urban environment using ultra low power sensors, wireless networks and web and mobile based applications. Sustainable smart cities are becoming a reality.

The maintenance of high quality of life in towns requires the innovative economic growth potential as well as our urban population in the urban areas should be stabilized at sufficient level in lieu with the resources available and the protection of environmental quality leading towards sustainable development. Zoning law is also required for the proper growth and development of a city.

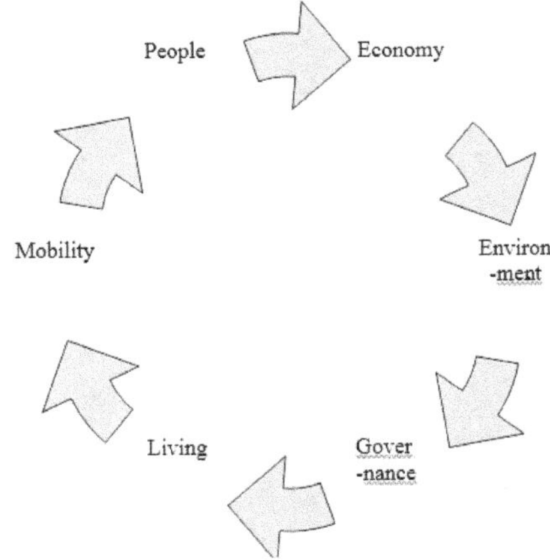


**Fig. 2.1 – Dimensions of Smart Cities**

# Chapter-4

# Consumerism and Environmental Degradation

Consumerism refers to the consumption of resources by people. The globe at present is in the midst of transition to a more consumption oriented life style. Early society consumed much less resources. With the onset of industrial revolution, consumerism has shown a rapid rise in few decades. It has been related both to population increase as well as increase in our demands due to our altered life style. In modern society, our necessities have multiplied which in turn resulted in the multiplication of consumption of resources. Globalization has brought technology to our doorsteps making communication, transportation and banking more accessible than ever before. For this reason, company executives have grabbed the opportunity to flash endless advertisements. Thousands of audiences have become the victim of these advertisers. Presently, the rate of consumption is rising at an alarming rate. The society in which we are living today is under threat of consumerism.

Consumerism is a social and economic order that encourages the acquisition of goods in ever increasing amounts. With industrial revolution especially in $20^{th}$ century, mass production led to over production beyond consumer demand, so manufactures turned to planned advertising to increase consumer spending. According to National Geographic, approximately 1.7 billion people world wide belongs to 'consumer classes'. They eat processed foods, desire bigger houses and cars and buy goods over and above their needs. While increased consumption facilitates jobs creation and stabilizes economy, the demand is stretching the planet's natural resources. Beside human beings, it is only environment which also suffers from the threat of consumerism. We are still not aware that we are contributing to environmental degradation through mass consumption promoted by advertisers. With increased consumption, our land, water and air get more polluted by industrial and tourism related wastes. According to United Nation, 86% of world's resources are consumed by only 20% of world's population. To meet their demands, limited resources are often transferred to luxury items instead of basic needs such as water, food, health and sanitation for poor.

In fact, if 20th and the present 21st centuries are marked by rapid socio- economic, scientific, technological and telecommunication advancements on one side, it is distressed by serious environmental issues on other side. Most of the environmental problems we see today can be linked to consumption said Grey Garder, Director of Research for World Watch. According to Diana Ivanova, a Ph.D. scholar at Noroegian University of Science and Technology, if we change our consumption habits, this would have a drastic effect on our environment footprint as well. She and her colleagues found that consumerism was much higher in rich countries than in poor countries and those with highest rates of consumerism had up to 5.5 times the environmental impact as the world average. Studies showed that stuff we consume – from food to knick- knacks- is responsible for up to 60% of global green house gas emission and between 50-80% of total land, material and water use.

Our consumption of goods obviously is a function of our modern culture. Only by producing and selling things and services does capitalism in its present form work and more that is produced and more that is purchased the more we have progress and prosperity. Due to increased consumer demands, the planet has been out of balance for many years. The present environmental crisis due to environmental degradation is the result of various types of pollution, depletion of natural resources, increasing energy consuming and ecologically damaging technologies, habitat loss due to industrial, urban and agricultural expansion, biodiversity loss  due to  excessive use of toxic pesticides, fertilizers, herbicides, deforestation etc. The current consumption patterns are depleting non- renewable resources, poisoning and degrading ecosystems and altering the natural cycle on which life depends. However, production, processing and consumption of commodities require extraction and use of natural resources (wood, ore, fossil fuel and water), it requires the development of factories whose operation creates toxic by products, while use of commodities themselves ( such as automobiles) creates pollutants and waste.

## Environmental Concerns

India, being a developing country, made progress in field of economic growth, technology, industrialization, urbanization, information technology and agriculture etc. The process of economic liberalization and globalization has resulted in making consumerism a way of life in middle class population of India. Our markets are flooded with various branded consumer commodities. Thus, middle class Indians are following the global trend of competitive and conspicuous consumption. But, our growth pattern and consumerist pattern have degraded the environment to a thinkable extent. Our main concerns are forest destruction, natural resources (water, mineral, sand, rocks) depletion, biodiversity loss, ecosystem loss, food and livelihood security for the poor etc.

Environmentalists point out three factors responsible for environmental pollution:- Population, Technology and Consumption. Population is the total number of people, consumption relates to amount each person consumes and technology determines how much resources are used and how much is being used/ wasted or pollution is produced for each unit of consumption. With environmental degradation, high levels of production and consumption require large inputs of energy. Additionally, it generates large waste quantities of by-products. Besides, the inordinate amount of waste that is generated by the consumer- oriented societies around the world has become a serious environmental issue. Most human activities are related to production and consumption cycles which produce excessive amounts of waste in form of solid, liquid and gaseous products. Excessive demand for consumer products has led to the creation of current environmental imbalances. As a result, these imbalances have caused an ecological disaster globally.

Society's addictive consumerism and environmental state- climate change, global warming is intricately linked. The industrial revolution is considered the 'go point' of climate change. Increased burning of fossil fuel (coal and petroleum) in thermal power plants, factories and vehicles etc. has increased the carbon dioxide concentration in the atmosphere which in turn intensifies Green House Effects resulting in rise of earth's temperature. This black carbon, methane and other air pollutants are emitted in large quantities in India every day from incomplete combustion of fuel, wood, crop waste and cattle dung etc. Facilitation of the growth of human population also led to increased pollution. More people equate to more waste and more carbon footprints, which leads to more pollution and global warming. Climate change and its effect will continue worsening, thus causing an environmental disaster.

Huge amount of waste is being produced by the urban industrial communities in form of plastic, paper, tin cans, bottles, thermocols disposable items, paper cups, pathological wastes from hospitals etc. Dead animals, agricultural wastes, animal excreta, fertilizers overuse etc. are rural wastes. As a result, we are polluting our environment in many ways. Millions of tones of wastes are poured in environment without proper treatment and disposal. Plastic garbage has created problems of littering in recent years. Plastic bags are also hazardous to environment since most of the people dispose them into waterways. Since plastic can be used as substitute for some forest products, the reuse and recycling of plastic is a necessary evil. With huge consumer demands, there are more vehicles that run on motor oil. As a result, oil pollutes the environment. It may end in a serious toll on future generations.

Sewage discharged from cities and towns is the predominant cause of water pollution in India. Open defecation is widespread even in urban areas of India. The tanneries of Kanpur city are heavily polluting the River Ganga by discharging of untreated water and causes deaths of aquatic ecosystems. Indian cities generate more than 100 million tones of solid wastes per year. Streets corners are piled with trash. India's garbage crisis is due to consumption. Solid waste landfills have become significant sources of greenhouse gas emissions and breeding sites for disease vectors such as flies, mosquitoes, rats, and other pests.

Forests are being depleted for various consumption and construction activities. Several luxury items such as refrigerators, air conditioners, dispensers, fire extinguishers release chlorofluorocarbons (CFC) into atmosphere. Similarly plastics, pesticides, paints, dyes, solvents etc. are more examples of chemical products made and disseminated for benefit of man. All these disturb the ecosystem. Therefore, we need to opt for growth patterns which are pro- environment, sustaining natural resources and creating less pollution and waste.

Besides, India is moving on path of economic development with continuous rise in middle class. This class use consumerism as a lifestyle with less concern for environment. Thus, we see that the trend of ever-expanding consumption puts immense strains on the environment. No doubt, consumerism is the outcome of our developmental process. People consume material products to make their life easier and stylish. But, the consumerism has a negative impact on our society as well as environment. The need of the hour is that environmentally sustainable pattern of production and consumption should be adopted. We have to control and reduce the exploitation of natural resources. Awareness should be spread about negative impact of using cosmetic goods, air conditioners and other modern appliances which release ozone depleting gases like chlorofluorocarbons. We should also check the production of wastes. Waste management should be based on 3 'R' ,principle i.e. Reduce, Reuse and Recycle. We are presently in a condition where earth's ecological state is unable to cope up with the prevailing pollution because of continuously increasing consumer demand.

We need to move to more sustainable consumption patterns that reduce environmental damage, improve efficiency in resource use and regenerate renewable resources such as water, air, soil, wood etc. Making eco-friendly and green choices in our daily life is as important as moving away from addictive minds that got us to such a non-environmentally friendly place.  To save the habitable planet, we need to think about our own ideas of how much is enough for today.

# Chapter- 5

# Society, Technology and Environment

When we think of environment, the things that immediately come to our mind are air, land, water, wildlife-animals, birds, fish and forests etc. However, the environment of man consists of both natural and manmade substances and conditions in which and by means of which human society satisfies its most varied material as well as cultural needs. In the society, man is connected with his environment by constant interaction. The environment affects man both physically and psychically. On the other hand, man constantly affects the material surrounding him, changing them (in ways both good and bad, from his own viewpoints and that of nature), and thus becoming to a great extent, the creature of the living conditions and style of the generations to come. They found environment to be a discouraging one for their rapid industrialization in the name of modernization. But due to this modernization, we have reached to a point where it is required to give a serious thought to the environmental problems resulting from this modernization. Every breakthrough in science and technology, one way or the other affects society, its social institutions and the surroundings i.e. environment. The most referred environmental problems are related to the earth's resources like soil, groundwater, forest, river, sea and wildlife. These problems largely relate to the conditions of poverty and under-developed vis-à-vis the negative effects of the very process of development in India.

The interrelationship and interdependence of human being and environment is not recent but dates back to Charles Darwin's theory of natural selection. Nature has been a part of Indians' social life. Environment and natural resources are God's gift to man and other living beings. There are symbiotic relationships between bio-physical ecosystems and social systems, with strong cultural interconnections between the two. His concept has related certain social characters with the survival of societal groups in natural environment. We people also share the same relation of ecological dependence with other inhabitants of the earth. But, they are different from other species because of having capability, culture and power to destroy, exploit and manipulate the natural environment according to their needs. Average Indian citizen is still very much dependent on his natural environment for satisfaction of his needs. Therefore, environmental degradation has become a threat for present society that how to cope up these eco-problems. Present day society is not able to produce any solution because it is disabled by values our leaders regularly and constantly speak, such as economic growth, jobs, consumerism, competition, power etc. The current society pursuing these aims cannot hesitate depleting their natural resources, producing much wastes and ultimately ruining the ecological balance as well as biodiversity. The answer goes only to 'social values' which have been expected to be fruitful in developing environmental ethics. Some of the traditional social values aimed at protection and preservation of ecological balance and environment.

Social values are associated with social life, religion, custom and psycho-spiritual aspects of people. Social values are cultural standard that contain a concern for other's welfare. These are representation of individual needs and desires on one hand while of social demands on the other hand. Social values enable individuals to work together to realize collectively desirable aims.

Environmental protection is everybody's concern but nobody's responsibility because wealth and power has dominated the modern society. In the past, man was considered as trustee of the environment, but today, in modern technological society, he considers himself as owner having instant gain and continuing unplanned exploitation of natural resources, ultimately ending with ecological degradation. I agree that 'Technology' has changed our lives for better and easier with activities uncomplicated. While on one hand making our lives easier, it has also had a devastating effect on humankind. Whether technology is more of boon or bane to our environment?

Technology was used to gear up the production of goods, better transport and delivery, methods of communication not only faster but also far reaching. Trade flourished fast due to technological advancement and brought the globalization. As the trade and commerce grow, the more it heightened, the people believed it to be turning point in society. But is this really making life profitable and relaxing? Human started creating objects that are harmful to environment and they called this as 'Technology'. It can be a powerful force to improve our standard of living, it comes at a cost. Technology does not operate in a vacuum; it does so in an environment, so the advancement of technology will surely influence the environment with all its components. There is a reciprocal relationship between environment and technology. The environment serves as a room or the place where the technology is put into practice. Therefore, it is inevitable that there will be a constant interaction between them. We are surviving in this planet because of well balanced ecosystems. The more we destroy the habitats of plants and animals, the more our ecosystem is going to disturb the balance. Environmental pollution, degradation and contamination of key components of natural environment - air, water and land is the result of technology mismanagement and lack of control measures.

Technology and society or technology and culture refer to cyclical co-dependence, co-influence, and co-production of technology and society upon the other (technology upon culture and vice versa). This synergistic relationship remained from dawn of humankind to invention of simple tools and continued into modern technology as printing press to present modern digital world. By 1980, use of technology started to escalate. It has entered and started playing major role in day to day life and the society. As society became aware the developing technologies, they started to take advantage of it. In fact, digital technology in the society has constructed another worldwide information and communication system in addition to its origin (computer etc.). With unparalleled power of internet to unite the whole world into global village, it has made communication easier more than any thing else. Digital technology is commonly used for downloading music and watching movies at home etc. Social media platforms such as instagram, facebook, snap chat and mobile smart phones have altered and revolutionized the way of life. These are only few of positive aspects of technology in society, there are also negative side effects on the environment and biodiversity.

At present, our life is wrapped under invisible field of electro magnetic radiations released by devices we use daily. Every person, plant and animal is affected by these radiations. WHO has said that problem will increase as the technology advances, but it is not advancing how can technology that is causing so much physical, mental and emotional disease be described as being an advance? Environmental degradation is a growing concern as continued industrialization being witnessed mostly in developing countries. As industrial activity increases requiring more raw materials, technology contributes towards depletion of resources such as coal, timber and wild animals. Farming activities such as burning of bushes, deforestation and usage of chemicals to enhance soil fertility is an environmental exploitive. Besides, expansion of tourism indiscriminately ends in devastating ecological damage as well as biodiversity loss. Nation continues to increase the use of technology in warfare and they produce weapons that make use of metals, chemicals and micro-organisms that have far greater negative effects. Large amount of mercury waste material was said to have been released in environment at time when nuclear weapons were being manufactured.

On brighter side, new green technology has come up with better solution in which the technology is environment friendly and is created and used in a way that conserves natural resources. It is believed that green technology promises to augments farm profitability while reducing environmental degradation and conserving natural resources.

Yes, technology definitely does have both pros and cons through line of distinction are becoming narrower by the day. The concept of value may play an effective role in maintaining an Eco-friendly society. There is need of environmental ethics so as to curb on part of everyone to restrict its want only to what is essential for human well being. Today, economic growth and environmental preservation should have to be balanced and harmonized. All the societies across the globe needs adoption of a strategic environmental management to improve the environment related issues within the nations. Therefore, the problems of "taking too much from the environment and restoring too little" need an intelligent and helpful solution. In finding a balance between technology development and environmental protection, a key principle is that one should attempt when possible, to find sustainable solution.

# Chapter-6

# Environment and Sustainable Tourism

The country like India is inhabited by people having different religions, languages, culture, and traditions. Tourism play an important role in maintaining harmony, national integration, love, peace and prosperity among Indians and abroad. Tourism and travel have been an important social activity of human beings. The term tourist has been derived from the Latin word *'Tornus'* which means a tool, a circle or turner's wheel. It means, a person undertaking a circular trip and ultimately comes back to the place from where he starts his journey.

Apart from Technology and Telecommunication governing the global economy in this 21ˢᵗ millennium, tourism also has emerged as fastest growing economic sector. Today tourism is the largest service industry in India with a contribution of 6.23% to the national GDP and 8.78% of total employment in India. An important feature of tourism is its contribution to national integration and creation of harmonious social and cultural environment. Its powerful economic and social force creates employment and valuable foreign exchange widening our understanding of society. According to an assessment, in India alone 100 million additional jobs will be created in next 25 years. India visits more than five million annual foreign tourist arrivals and 562 million domestic tourism visits. There has been a rate race among developed and developing countries to expand tourism. The ministry of tourism is a nodal agency for development and promotion of tourism in India and maintains the "Incredible India Campaign". India is ranked 14th best tourist destination for its natural resources and 24th for its cultural resources with many World Heritage Sites, both natural and cultural, rich fauna and flora and strong creative industries in the country.

Tourism Industry in India has several positive and negative impacts on society and economy. Generating income and employment, foreign exchange earning, promoting peace and stability are the positive impacts while the negative impacts are undesirable social and cultural change, increase tension and hostility and adverse effect on ecology and environment.

## Impact of Tourism on Environment

There exists a crucial interdependence between tourism, the environment and community adjacent and within tourism destination. The key resources for the most popular tourist destinations is natural environment, coastal resorts, tropical rainforests, wildlife in natural parks and alpine ski resorts, all rely on a mixture of natural beauty, good weather and safe conditions to attract holiday destination being landscape and natural environment, followed by climate, cost of journey and historical features of the place to visit. Tourism has turned biodiversity into one of its most marketable assets. Images of natural environment, unspoiled rainforests, crystal clear waters and wild life in their natural habitats are a standard fare in tourism advertising program such as "Incredible India". The areas that are appealing to tourists are often places with rich biodiversity.

Tourism industry in India can have several positive impacts on the environment and biodiversity.

1. ***Direct Financial Contribution:*** Revenues from park and zoo entrance fees and similar sources can be allocated specially to pay for the management of environmental sensitive areas.

2. ***Government Revenues:*** Department also collects money in indirect ways that are not directly linked to parks or conservation areas. User fees, income tax, license fees can provide government with the funds needed to manage natural resources.

3. ***Raising Environmental Awareness:*** Tourism has potential to increase public appreciation of the environment and to spread awareness for environment.

4. ***Improved Management:*** Because of their attractiveness, natural resources are identified as valuable and need to keep attraction alive can lead to creation of natural parks, wildlife parks, forest preservation etc.

Besides many positive impacts, tourism has some adverse/negative effects on environment. Though biodiversity works like a magnet for tourism, yet it is also one of the greatest victims. Saturation mass tourism with its impact on the environment does not help biodiversity contributing to the disruption of the delicate balance within ecosystem and in turn to the irreversible loss of life forms. Besides, its positive impact, the expansion of tourism indiscriminately results in severe ecological damage to host country as well as threaten the biological resources on which economic activities depend. The tourism consumes significant amount of natural resources and can degrade ecosystems, may raise the cost of living for local people, may degrade local culture and sell it as a 'commodity', and its revenues may flow out of the destination with few local benefits. Developmental activities have taken a toll on environment. One of the major causes of degeneration of environment is construction of infrastructure disturbing the flora and fauna. It is paradox that while needing biodiversity to flourish, tourism allows its degradation. The problem is compounded by the fact that tourism often occurs in areas of high biodiversity, such as coastal zones, mountains and protected areas. Major cause of degeneration of environment is unplanned and unchecked activities in tourist places.

Three main impacts areas of tourism are depletion of natural resources, pollution, and physical impacts: -

## 1. Depletion of Natural Resources

Tourism development can put pressure on natural resources when it increases consumption in areas where resources are already scared. Tourism industry generally overuses water resources for hotels, swimming pools, golf courses and personal use of water by tourists resulting in degradation in water supplies. This can result in water shortages and water supplies, as well as generating a greater volume of waste water. In dryer regions like Rajasthan, the issue of water scarcity is of particular concern. It also creates scarcity of food, energy and other raw materials. Increased construction for tourism and recreational facilities has increased the pressure on natural resources and on scenic landscapes. Direct impact on natural resources both renewable and non renewable in provision of tourist facilities can be caused by use of land for accommodation and other infrastructure provision and use of building materials. Forests also suffer deforestation caused by fuel wood collection and land cleaning for construction of hotels and lodges. Trekking activities in mountains often result in loss of flora. Uncontrolled mass tourism is one of the main drivers of coastal degradation.

## 2. Pollution

The increased pollution is another impact of tourism on biodiversity. Tourism can cause the same forms of pollution as any other industry, air emissions, noise, solid waste and littering, releases of sewage, oil and chemicals, even architectural/visual pollution. Transport by air, roads and rail is increasing day by day in response to rising number of tourist activities in India. Tour buses also cause air pollution as they emit CFC, Carbon dioxide and other green house gases. These are linked with acid rain, global warming, climate change consequently resulting in some ecosystems to exceed critical threshold. Some of these impacts are quite specific to tourist activities where the sites are in remote areas like Ajanta and Ellora temples. For example, tour buses often leave their motors running for hours while the tourists go out for an excursion because they want to return to a comfortably air-conditioned bus. Noise generated by powerful speed boats, cars, buses as well as recreational vehicles may scare animals in the area and affect creatures living below the surface affecting their natural activity and breeding pattern. In addition to causing annoyance, stress, and even hearing loss for humans, it causes distress to wildlife, especially in sensitive areas.

Water itself is likely to become polluted with various chemicals, paints on the underline of boats start peeling off after time and leaves a dangerous oxide in the water that reduces oxygen level. The oil and petrol of recreational vehicles are harmful to aquatic wildlife like coral reefs. A dramatic rise in temperature of water by global warming has been causing a large scale bleaching process destroying entire coral population due to  vanish of algae living in coral tissues.

Waste disposal and improper disposal can be a major despoiler of natural environment – scenic areas, oceans etc. In mountain areas of the Himalayas and Darjeeling, trekking tourists generate a great deal of waste. Tourists on expedition leave behind their garbage, oxygen cylinders and even camping equipment. Such practices degrade the environment particularly in remote areas. Wastewater has polluted seas and lakes surrounding tourist attractions, damaging the flora and fauna. Sewage runoff causes serious damage to coral reefs because it stimulates the growth of algae, which cover the filter-feeding corals, hindering their ability to survive and damage marine ecosystem. Besides, coral reefs- one of the marine world's natural treasure- are facing bleaching which dramatically whitened. This damage would wreak havoc in fisheries and tourism, disrupting the economies of many nations. Changes in salinity and siltation can have wide-ranging impacts on coastal environments and sewage pollution can threaten the health of humans and animals. Examples of such pollution can be seen in the coastal states of Goa, Kerala, Maharashtra, and Tamil Nadu, etc. The Dal Lake (Srinagar) which was once pristine has lost its nature due to the tourist pressure and is now covered with animal carcasses, sewage and weeds. The lake has shrunk as it was unable to handle the pollution caused by regular tourist influx.

Golf courses have brought heavy ecological and social costs: deforestation, biodiversity destruction and erosion, dispossession of residential homes and farms, over consumption of water, very high use of pesticides and fertilisers which threaten local residents, workers, wildlife and golfers themselves leading to suffer from various skin and respiratory diseases. Besides, due to depleted water supply to golf courses, agricultural production come to halt and farmers are forced to migrate to urban areas for employment.

Tourism also resulted in disorderly and scattered tourist facilities which generally are not eco – friendly and leads to aesthetic degradation of landscapes.

## 3. Physical Impacts

Attractive landscape sites such as sandy beaches, lakes, riversides and mountain tops and slopes are often traditional zones characterized by species rich ecosystem. Typical physical impacts include–

- ☆ Construction activities and infrastructure development for tourist facilities can involve sand mining, beach sand dune erosion and lead to land degradation and loss of wild life habitats. The area of concern which emerged at Jaisalmer is regarding the deterioration of the desert ecology due to increased tourist activities in the desert.

- ☆ Deforestation and unsustainable use of land.

- ☆ Marina development and break waters can cause changes in currents and coastlines resulting in disruption of land sea connection, for example sea turtle nesting spots, coral reefs etc.

- ☆ Trampling – Tourist using the same trail over and over again trample vegetation and soil causing loss of biodiversity.

☆ Anchoring of boats, sport fishing, scuba diving, yatching and cruising etc. can cause direct degradation of marine ecosystem. Examples may be cited from Krushedei Island near Rameswaram. What was once called paradise for marine biologists has been abandoned due to massive destruction of coral and other marine life.

☆ Wildlife viewing can bring about stress for animals and alter their natural behavior as tourists chase animals in their trucks. Safaris and wildlife watching activities have a degrading effect on habitat as they often are accompanied by the noise and commotion created by tourists.

☆ Climate change might even cause some ecosystem to exceed critical threshold leading to their irreversible decline.

☆ Export for decorative purposes - Commercial exploitation of aquarium fishes, coral reefs, orchids, marine turtles leads to face danger of survival in the wild. Orchid's flora is now endangered.

It is definitely true to say that majority of global tourism has negative impact on environment and biodiversity. If we continue to exploit the environment at rate that we are doing, soon there will relatively little left to enjoy. A WWF study published on " Climate Change and its impact on tourism" warned that droughts, rising seas, flash floods, forest fires and diseases could turn profitable destinations into holiday horror stories. Thus, biodiversity – the library of life is now threatened. So, if we want to enjoy nature one must preserve it otherwise all the exotic destination will become extinct. Eco-friendly tourism should be promoted all over the world.

## Eco-tourism

It is a responsible travel to natural areas that conserves the environment and improves well being of local people. Responsible tourism is a new concept in the tourism industry and developed by Jost Krippendorf in 1980's. He has an aim to "to develop and promote new form of tourism, which will bring the greatest possible benefit to all the participants- travelers, the host population and the tourist business, without causing intolerable ecological and social damage". In a nut shell, responsible tourism is that form of tourism in which initiatives and responsibilities are taken by both the tourist and the travel agencies to maximize the positive impacts and minimize the negative impacts on environment, traditions and cultural heritage of the region. According to policy and guidelines of Govt. of India 2002), " Eco- tourism means sustainable, equitable community based endeavor for improving living standards of indigenous host communities". However, World Trade Organization (WTO, 2004) defines," eco-tourism as tourism that involves travelling to relatively undisturbed natural areas with specific objectives of studying, admiring and enjoying the sceneries and its plants and animal, as well as in existing cultural aspects found in these areas". Historical, biological and cultural conservation, preservation and sustainable development are some of fields closely related with eco-tourism. Eco-tourism differs from 'resort tourism' or 'mass tourism' by requiring lesser infra structure development and lower impact on the environment. There are three specific segments of eco-tourism:

1.  Nature /green tourism: based on the web of life or life forms.
2.  Adventure tourism: dealing with sports activities in various natural environs.
3.  Culture tourism: dealing with aspects of social and cultural heritage.

Eco-tourism as a developmental tool and also a tool for biodiversity planning and conservation should be emphasized. The potential of eco-tourism as an economic activity will contribute to the sustainable economic and social development of the country.  It is rapidly growing global trend. Some of the nations like South Africa, the United Kingdom, United States, Sri Lanka and India are already practicing eco-tourism and gradually other nations are also following the ideologies of these countries.  In the upcoming years, India will become a hot tourist destination of South East Asia. In India, Kerala has pioneered in practicing innovative tourism ideas. It is a state which is rich in tradition and culture. In Kerala where people live in houses made in indigenous style. It has started eco-tourism in the state to create awareness among all stakeholders for a better environment to live and visit.  North East states have also vast resources for eco-tourism. Indigenous tribes of North Eastern region lead an agrarian life style. Himalayan states in India are also a major destination for eco-tourism. The Sunderbans, the world's largest mangrove forests represent a unique paradise for nature lovers and of eco tourists. " Sunderban Jungle Camp" is the project of India initiated by Help Tourism group to connect the heritage sites and protected areas of Sunderbans with the livelihood of the local communities through community based, environmentally and socially responsible tourism. Another successful eco- tourism project is "Awake and Shine" started by retired Indian Army General- Jimmy Singh in May 2006 to develop the Samthar Plateau near Kalimpong as a tourist destination depending on the home stay holidays, local handicrafts and other products. It helped the local communities to improve their lifestyle. Another project of India is in the Anakkara Spice Tourism Village in Idukki district of Kerala, started in October 2004 by Women in Agriculture (WIA), a group of women. WIA has joined hands to preserve nature, share benefits of tourism among local people and boost the local economy. It has made these ladies to improve their living as well as their family well being.

## Sustainable Tourism

It is responsible tourism intending to generate employment and income along with alleviating any deeper impact on environment and local culture. Though the goals of eco-tourism and sustainable tourism are much similar but the latter is broader and conceals within itself very many aspects and categories of tourism. Major challenge is, therefore, to enhance economic benefits of tourism while limiting its negative environmental and social impacts. The most widely accepted definition of sustainable tourism is that of the World Tourism Organization. They define it as "tourism which leads to management of all resources in such a way that economic, social and aesthetic needs can be fulfilled while maintaining cultural integrity, essential ecological processes, biological diversity and life support systems". In addition they describe the development of sustainable tourism as a process which meets the needs of present tourists and host communities whilst protecting and enhancing needs in the future. The importance of sustainable tourism lies in its motive to conserve the resources and enhance the value of local tradition and culture.

The sustainable tourism may result in significant positive impacts for biodiversity conservation with social benefits to host communities. Various positive impacts are-

- ☆ Raising awareness amongst tourists to promote conservation.

- ☆ Sustainable land management- by providing alternative forms of livelihood for farmers and rural communities which are dependent only on natural resources.

- ☆ Providing cultural appreciation- It provides incentives for maintaining traditional art and crafts, traditional knowledge and practices that help to sustainable use of biodiversity.

- ☆ Economic value for habitat protection- Tourism already makes a contribution to income for protected areas and other attractions through entry fees, permits etc. which can be invested in programmes for local communities to manage protected areas.

- ☆ Rising revenue for local communities- Provide opportunities for job creation and business development.

Sustainable tourism is thus, viewed from four related aspects-Economic Sustainability, Ecological Sustainability, Cultural Sustainability and Local Sustainability.

Sustainable tourism requires strategic planning, policy framework, a legislative framework, collaboration with local host communities, the private sector, and individuals. Any development of tourism requires knowledge, vision, commitment and training to make them democratically driven, culturally legitimate and socially sustainable. Sustainable tourism planning process for integrated tourism and biodiversity planning is shown in figure- 3.1 and figure -3.2.

In 1994, World Travel And Tourism Corporation(WTTC) initiated the 'Green Globe' , an Agenda 21 based industry improvement programme, which provides guidance material and a certification process linked to both ISO standards and Agenda 21 principle. 'Green Globe' has also developed a specific destination program, which provides a methodology for Travel and Tourism destinations to implement sustainable development. The aim of 'Green Globe' is to become the primary global standard of environmental commitment by the global tourism industry.

## Pro-Poor Tourism

There are a number of reasons to look again at tourism and to assess its potential to generate pro-poor growth. The natural resources and the local culture are amongst the few assets of the poor. Pro-Poor tourism generates net benefits for the poor. It may be defined as forms of tourism where the benefits to the poor are greater than costs which tourism brings them. Pro-Poor tourism aims to expand opportunities for those living on less means. Tourism can affect many of these, positively and negatively, often indirectly. The application of a 'sustainable' livelihood framework is essential to develop pro-poor approaches. Pro-poor tourism should not just be pursued in nitch markets( such as eco-tourism or community tourism). It is even more important that mass tourism developed in ways that benefit the poor. It is time to consider the role of tourism in contributing to pro-poor development.

Tourism should be judged against other possible strategies and where it offers the best opportunities for pro-poor growth or where it can make a useful contribution by increasing the diversity of opportunities for the poor.

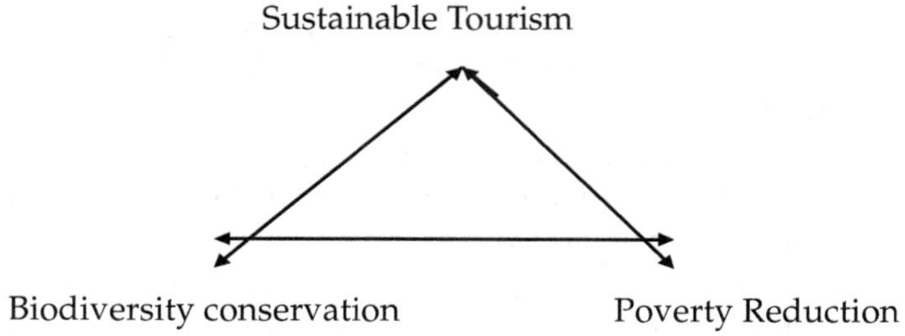

Sustainable Tourism

Biodiversity conservation                                    Poverty Reduction

**Fig. - 3.1**

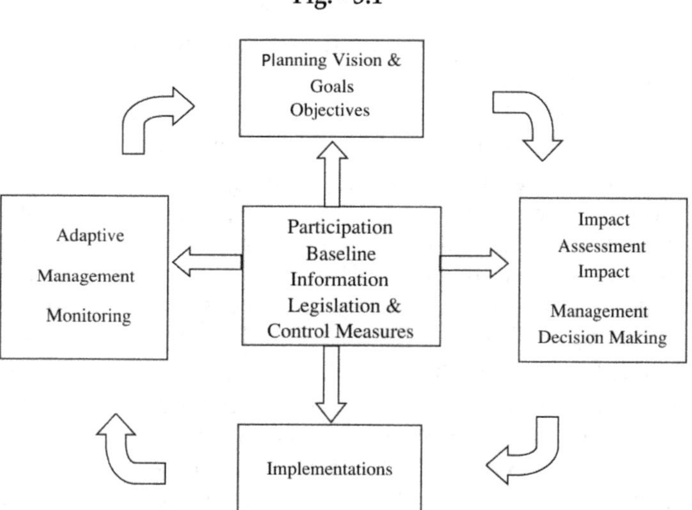


**Fig. 3.2 : The Planning Cycle for Integrated Tourism and Biodiversity Planning**

Tourism industry in India is growing and it has vast potential for generating employment and earning large amount of foreign exchange besides giving economic and social development. Eco-tourism needs to be promoted as a strategy that can protect 'Natural Wonders' at the same time that it contributes to the welfare of the people so that indirectly helping in preserving and sustaining the diversity of the India's natural and cultural environments. Tourism in India should be developed in such a way that it accommodates and entertains visitors in a way that is minimally intrusive or destructive to the environment. Government and tourism industry must abide to the principle that environmental protection is an integral part of tourist development. Only if tourism investors and developers consider the natural capacity for regeneration and future productivity of natural resources; and accept that people, communities, customs and lifestyles have an equitable share in the economic benefits of tourism; and recognise the local people in tourist destinations, tourism may become sustainable. Awareness programmes and education are, therefore, imperative. There is dire need to felicitate and encourage dialogue between conservation interests and tourism industry about the importance and fragile nature of ecosystem, heritage, and cultures to achieve a sustainable future for them. Therefore, conservation of biodiversity as well as sustainable use of natural resources can and must be incorporated into tourism development policies that bring economic and social benefits to host communities. Hence, conserving the ecological integrity and environment is imperative if tourism is to be sustained.

# Chapter-7

# Global Warming:
# A Byproduct of Globalization

Globalization can be described as a process by which the people of the world are unified into a single society and function together. This process is a combination of economic, technological, socio-cultural and political forces. It refers to the increasing integration and interdependence of countries shaping world affairs around the globe. Process of globalization is popularly described as a gradual removal of barriers to trade and investment between nations. It aims to achieve economic efficiency through competitiveness, while seeking broader objectives of economic and social developments. Globalization has led to faster access to technology, improved communication and innovation. Never before has been so likely that fruits we eat, shirts we wear, cars we drive and movies we see were produced in other countries. Globalization is a broad term that encompasses the increased global interdependence in the economic, social, technological, cultural, political and ecological spheres. It is an ongoing process and its basic character is exploitative, coercive and domination. It has important implications for environmental challenges such as deforestation, climate change, pollution, biodiversity loss, depletion of natural resources and water crisis etc. Environmental effects of globalization are particularly concerned with the question whether a world wide liberalization of trade may provoke environmental collapse. In relation to trade, there are three major environmental concerns- Domestic environmental effects caused by use of imported products; foreign environmental effects caused by production of exported goods; environmental effects caused by transport movement needed for international trade. Trade should be a tool to share human aspirations, to improve living standards and quality of human life, but it should not allow to degrade world or trade of most valuable things like clean air, clean water, wild life and wild places. Changes in international trade patterns, markets, technologies and communication patterns affect both the economy and environment. A demanding problem of globalization is formed by environmental decay caused by rise in international transportation. Transport contributes to air and noise pollution, intrusion to land scape, congestion etc. It is contributing to main global problems- green house effect and ozone layer depletion.

Globalization has allowed industrialization to spread further across the globe. Rapid industrialization and soaring human population have led to a massive imbalance of natural ecosystem. It is altering Global environment. Some have an opinion that ecological impact of globalization is positive, as a force of progress. It fosters economic growth. Others perceive the impact as negative, as a force of ecological decay. But it is a destructive process consuming too many natural resources.

## What is Relation Between Globalization and Global Warming?

Globalization and Global Warming both have 'global' in their name, which means that they affect entire world. There is also an indirect connection that globalization helps many third world countries to industrialize and develop, and that enrich billions around the world with that level of economic and industrial development we are looking at increased consumption of fossil fuel and other causes of global warming, for example- China has developed from a nation of bicycle to a nation of cars. The smog in Beijing was not a thing before economic development.

Globalization has a huge benefit in raising the standards of living for most of the world. It is best time for a person to be alive ever and continues to improve day by day. The globalization processes influence human behaviour in various ways and apart from positive effect, they have a differentiated effect on environment. Pollution has become need of the developed countries to enjoy comfort. But, without admitting it, they are 'helping' the developing countries to develop by the process of globalization. Now, one of the side effects of globalization is global warming.

## Globalization and the Environment

Globalization and environment are intriguely interconnected. The relationship between globalization and environment is too complex and they are inter- related to each other at every level in the present scenario. The globalization has put added pressure on environment, for example- World Bank figures show that deforestation for cropping accounts for up to 20% of world's global Carbon emissions. Increasing trade has also encouraged fishing, destruction of forests and the spread of pollution to previously unrecorded levels.

With the development of third world countries into improved industrial regions, there is threat of increased emissions both from vehicles and factories. This process of industrial production is either harmful to the environment or to workers or both and continue to harm the environment and contribute to global warming. Another way is via logging. Trees are the 'Lungs of the Globe' and their ability to regulate carbon dioxide is well documented. Trees are cut down and semi-processed in developing nations and shipped to developed nations who have already killed most of their trees. It can be said that the increase in cutting down of trees was due to globalization.

The natural resources are not inexhaustible. Oil is fast depleting. Nearly 1/3 of world's population is starved for water. Global population has tripled in past 70 years while water use has grown six fold due to industrial development, wide spread irrigation and lack of conservation. It is feared, scarcity of water may lead to third world war. There is a projected 3° C jump in global temperature leading to global warming which in turn would include a loss of up to 400 million tons of cereal production and put half of the current world population at risk of water. It is a wake up call for the developed industrial nations.

The increasing pattern of globalization has also put forth consequent effect on our ecosystem. Industries emitted various Green House Gases which acted as a major contributor in climate change. Our atmosphere is gradually warming due to accumulation of these green house gases. Global Warming also called 'Climate Change' refers to worldwide rise in temperature as a result of 'green house effect'. Climate change is an issue that affects all society on the planet. Global Warming is a result of global agriculture and industrial processes, all of which serve to meet the needs of the global market demands. The consequences are also global with all regions of the world experiencing some form of change as a result of global warming. Impacts are also not evenly distributed, for example- Bangladesh which not equally contributing largely to global emission will sustain problem of flooding from sea level rise. Other implications of global environmental problem are acid rain, dumping of waste in international waters, depleting of climate ozone, and loss of species, biodiversity and resources etc. Due to globalization and industrialization, various chemicals have been thrown into soil which has resulted into the growth of many noxious weeds. Toxic wastes have caused a lot of damage to plants by interfering in their genetic makeup. Globalization also creates a demand for genetically modified products. Another argument against globalization is misuse of resources due to increased raw material trade. In various parts of the world, mountains are being cut to make way for a passing tunnel or a highway. All the countries are linking with each other for further exploitation of the resources. Resources are limited and also limited to country specific, for example- New Zealand is rich in mineral resources whereas Asian countries are enriched with human resources. So, as a whole, countries compete each other globally. Developed nations would use their economic strength to extract natural resources from third world countries where by the time, resources would become extinguished. Without some form of regulation in form of international carbon taxation, globalization would fuel global warming and increase the rate of climate change.

## Global Warming and Human Activities

Greenhouse gases are released by activities like burning of fossil fuels, land cleaning and agriculture. The problems begin when human activities accelerate the natural process by creating more green house gases in the atmosphere than are necessary to warm the planet to an ideal temperature.

| | | |
|---|---|---|
| ☆ | Burning of natural gas, coal and oil- | Raises level of carbon dioxide which is main culprit of Global Warming. |
| ☆ | Agricultural practices- | Increase the level of methane and nitrous oxide. |
| ☆ | Industries- | Produce long lasting gases which destroy the ozone in earth's upper |

atmosphere and enhance green house effect and global warming.

☆ Deforestation-

Trees use carbon dioxide and release oxygen which create optimum balance of gases in the atmosphere. But forests are logged for timber or cut down for urbanization and farming.

☆ Population-

Enhances global warming as more people use fuel for heat, transport etc. which increases green house gases.

☆ Use of pesticides and insecticides-

Extensive use of DDT and other pesticides had many effects other than eradicating unwanted pests, it affect the entire Food Chain including greatly depleting the songbird population.

☆ Use of non-degradable plastic-

Caused the remotest beaches with tons of plastic waste.

☆ Water crisis-

Over use of ground water for irrigation has depleted underground aquifers.

☆ Habitat destruction-

Comes from land development resulting many plants and animals to become extinct because multinational corporation select spots where raw material, labor and regulation are cheapest.

☆ Tourism-

Global tourism expansion indiscriminately results in severe ecological damage to host country as well as threaten the biological resources.

Human activities, since the industrial revolution, have increased the content of gases in the atmosphere that trap solar heat. At the same time, human activities have reduced the forest cover that had traditionally absorbed this same material. An increasing level of these atmospheric constituents has lead to a chain of chemical and physical changes that have increased the world's temperature. This global warming leads to increased melting of ice covering both poles, thereby resulting in sea level rise. Sea level rise has been called "The dipstick of climate change". These changes may increase the frequency of change weather events such as, floods, droughts, heat waves, hurricanes etc. According to Tara Buakamsri of Greenpeace, Bangkok, at present, is sinking one to two centimeters a year and there is a risk of massive flooding in near future. Seasonal variations affect agricultural and animal reproductive patterns worldwide. During recent years, tropical sea surface temperature has begun highest causing the coral reef die off. There can be no longer by any reasonable doubt that global warming -the world is experiencing today, is a side effect of fossil fuel use. A clear correlation exists between industrialization over past 150 years and rising green house gases. A major climate change shift is underway that could have a severe impact on the way we live.

The Global Warming we are experiencing today is a byproduct of globalization. Explosive growth of global economy threatens the natural system that sustains life on earth. Besides increasing efficiency, globalization is devastating natural habitats, spreading global warming and increased air and water pollutions. Due to increasing global nature of trade and business, traditional national environmental protection techniques have become less effective. Both developed and developing countries have been satisfying their needs in terms of globalization. Decades ago, Mahatma Gandhi said *"the earth provides enough to satisfy man's need, but not any one's greed"*.

Globalization is neither globally beneficial nor environmentally sustainable. Globalization is driven by western culture enriched in debt and oblivious to responsible, sustainable living. Global Corporate rely on transforming agricultural societies into exploitable industrial culture. Globalist profit from shifting wealth from country to country in a never-ending quest for cheap labour and resources. Their wealth comes at the expense of the environment. * Think* Globally; *Act* locally. Localization is key: A return to self sustaining Communities and Nations will ensure steady, responsible economic growth for all. Therefore, proper balance of globalization and environment is badly needed.

The earth that contains abundance of life – is a vast system. Therefore, our actions and interventions must be informed by systematic thinking, with the realization that each and every action has an effect on the system as a whole. We must learn to pay attention to the costs and consequences of our actions so that it is possible to choose those that are most environmentally conscious and beneficial. Technological solution will determine the future of globalization as well as the global environment. But they will do so within the context of global consumption demands. Technology can't change the demands or help us, satisfy all of them but it can through globalization, help and meet these demands in a more planet – friendly.

# Chapter-8

# Climate Change and Environment

All living organisms including us, the human beings, and environment are intimately interconnected interacting mutually with each other. Earth system is not, and never has been free from change. Natural variability is an intrinsic part of all environmental systems, and only the short time span of direct human observation gives a mistaken impression to think that earth environment has not changed at all! Climate on earth has changed on all time scales, including long before human activity could have played a role. Changes in earth's radiation balance were the principal driver of past climate changes, but the causes of such changes are varied. The earth climate especially the mean global temperature has been changing at different time scales and each such change has direct relationship to the biosphere in terms of species level changes or small scale terrestrial changes in response to variations in micro climatic factors. The hydrosphere has responded to changes from continental to marine environment. For each case- be it the Ice Ages, the warmth at the time of the dinosaurs or the fluctuations of the past millennium- the specific causes must be established individually. The record of past changes provides us with clues about geomorphologic and biological responses to future climate change, particularly by observing how these Earth surface systems have behaved during previous warming phases such as that at the end of the last glaciations. Early post glacial migration indicate that most biota (like trees) take centuries to adjust to a substantial rise, although some groups (such as insects and weeds) have been effectively able to adjust with any climate change. Therefore, response to warming was strongly individualistic by species. Ultimately, familiar communities of flora and fauna would break up in a greenhouse world and new species mixtures would emerge. It may be a modification of Darwin's theory of evolution by natural selection – in the present cases the survival will be only for those who can thrive changes in the environment brought about by man made activities such as green house gases.

Climate change is a phenomenon which is triggered by the change in average temperature of the earth. Changes in climate occur as a result of both internal variability within climatic system and external factors, both natural and anthropogenic. Although records of climate changes of earth are available for about the past 1000 years in Europe, and somewhat longer in Asia such as India and china. The earth has already witnessed seven cycles of glacial advance and retreat in last 6,50,000 years with sudden end of ice age around 10,000 years ago which also marked the beginning of modern human civilization. The climate change that our earth is witnessing and going to observe in coming future is different from the past incidences. All previous changes were due to some natural climatic processes but we, the human beings are the main culprits of present day climatic change. During the post industrial revolution especially after early 20th century, human activities have ignited to long lasting and irreversible changes in earth atmosphere. Present climate change is mainly caused by anthropogenic emissions of green house gases like carbon dioxide, methane, nitrous oxide, chlorofluorocarbons etc. The increase in concentration of these gases has enhanced the global average temperature. The industrial activity have raised the atmospheric carbon dioxide concentration from 280 parts per million (ppm) to more than 400 parts per million (ppm) in last 150 years.  It is released in the atmosphere through various natural processes such as respiration and volcanic eruptions and also through human activities such as change in land use pattern, deforestation, burning of fossil fuels etc. It is regarded as the most important and long lived force behind climate change.

## Current Scenario of Climate Change

As per the report of Intergovernmental Panel on Climate Change (IPCC), over the last 50 years, rise in temperature was 0.1°C per decade and expected will be 0.2°C per decade (IPCC, 2007). The same panel (2013) also forecasted a temperature rise of 1.4°C to 5.8°C over the next century i.e. year 2100. If these changes are affecting or going to affect the environment, then some evident and unprecedented changes should appear in various fundamental environmental components.

Because of the combined effect of greenhouse gases, the changes that are happening now are certainly rapid enough, leading to rise in average global temperature. Such increase will make the globe hotter than it has been in earlier years. There will also be more warming in winter than summer.  The number of droughts, storms, floods, heat waves and other extreme events are on rise too. The driving forces of climate change act in a mysteriously complex way. Humans adjust well to these changing conditions using modern technologies and equipments, but animals and plants are still struggling for adjustment in such drastic changes. Because of climate changes, species may no longer be adapted to the set of environmental conditions in a given region and could therefore, fall outside its climate nitche. Climate change is one of the most important global environmental challenges affecting all natural ecosystems. Climate and natural ecosystems are closely related and dependent on each other and the stability of this relation is an important ecosystem service. To estimate the effect of climate change on species, scientists use what they call a climatic envelope (sometimes also as a bioclimatic envelope), which is the range of temperatures, rainfall and other climatic-related parameters in which a species currently exists. As the climate warms, the geographic location of climatic envelopes will shifts significantly, possibly even to the extent that species that can no longer survive in their current locations. These species will need to follow their climatic envelopes by migrating to cooler and moister environments. Harsh conditions promote the shifting of vegetation from high altitude to low altitude. Complete eradications are generally noticed in area of human induced climate change such as eutrophication in water bodies, acid rain near industrial area, destruction of forests for real estate developments etc.

The long Indian coastline is also under threat. A recent study has detected rising sea levels in parts of the Indian Ocean, including the coastlines of the Bay of Bengal, the Arabian Sea, Sri Lanka, Sumatra and Java. The sea level rise, which may aggravate monsoon flooding in Bangladesh and India, could affect hundreds of millions of people inhabiting the coastline. Even minimal rise in sea levels will directly affect the world's biggest and most densely populated urban centers with some facing complete submersion. For instance, Lakshadweep, tiniest union territory of India, has 36 coral islands. Similarly, The Andaman and Nicobar islands are blessed with unique tropical rainforests, made up of mixed flora of about 2200 varieties of plants. Besides, islands are rich in ecosystem diversity too ranging from mountain forests to coastal wetlands.

Glaciers are melting almost everywhere around the world including Alps, Himalayas, Alaska and Africa. Anarctica is also changing. Dr. Felix Bast of Central University of Punjab, Bhatinda found green growth of moss at several patches on the islands and oases. He also found pink ice algae in Antarctica glaciers. The massive bloom of Clamydomonas nivalis- the pink algae was covering the large areas of glaciers. This algae is known to cause glacial melting directly and indirectly- directly, it melts ice in order to complete its life cycle; and indirectly, it masks the white colour of snow and therefore, reduces the albedo effect ( reflecting power of white surfaces). When white colour is masked by pink, the energy from sunlight gets trapped, enhancing the melting of ice. Laminaria (commonly known as Kelp), an invasive seaweed is also reported throughout the coastlines of several islands, especially Twin Island. The short term effect of changing climate is predicted to be an Anarctica that is getting 'greener'. The long term effect is bleaker, collapse of Anarctic ice sheet and the continent. It is estimated that up to 90% of increased atmospheric temperature gets trapped in the world's oceans. Oceans get heated up particularly in the tropics and salty warm water travels pole-wards. Warm water reaching the poles directly melts the floating ice there ( sea ice and ice shelves). As per current melting trend of Gangotri Galcier, it is expected to become extinct in next 2000 years.

Natural hazards are purely natural but changing climate conditions may exacerbate them. Heavy rainfall with subsequent increase in flood peaks result in catastrophic events. The extreme weather pattern increases the intensity and frequency of hydro- meteorological disasters. Climate change is a threat multiplier for instability in fragile Himalayan region. One of the most severe instances of this phenomenon is the flash flood along with debris flow at Kedarnath in June 2013, which has resulted in more than a thousand casualties. The Indian Himalayan belt is highly prone to all kinds of natural hazards like landslides, earthquakes, avalanches, flash floods and glacial lake outbursts etc. Evidences of global warming in Himalayas are declining snowfall, retreating glaciers, drying up of perennial springs, disruption in rainfall pattern and river flooding. The future climate would be wetter and warmer than the present one which will affect the flow of river, amount of snow, rainfall pattern in the region and flood peak pattern. Changes in glaciers and snow cover pattern can be considered as the direct indicators of rise in atmospheric temperature. Increased melting of glaciers in coming years may influence water flows in Himalayan rivers.

Nature has been reacting to climate change by altering behaviour and movement. For example, with the winters getting warmer, flower change their flowering period and owls develop darker body colour. It can have serious impact on its crops, forests etc., which in turn affect the national developmental goals such as poverty, health, employment and education. Assuming a global temperature rise of 4.4°C by 2080 over the cultivated areas, India's agricultural output is projected to fall by 30-40% which would be quite alarming unless proper remedial measures are taken. We might witness frequent crop failure and severe pest attacks on crops which may create a famine like condition. Emergence of different new pathogenic viruses and bacteria may worsen the situation. Due to warmer weather, soil erosion and soil degradation are more likely to occur. Climate change is likely to have a variety of impacts on soil quality and fertility. The decrease in stratospheric ozone may affect agriculture adversely because increase in dangerous Ultra Violet B can directly affect plant physiology and cause massive mutations, and indirectly through changed pollinator behavior. All these result in lower yield.

Due to fragile environment, mountain ecosystems are the most vulnerable to impact of climate change. The North eastern states of India are expected to be greatly affected by climate change because of their geo-ecological fragility, international borders and their Trans boundary river basins. Combined with rising temperature, this would lead to significant reduction in species abundance in this region. Invasive species like *Lantana camera, Parthenium hysterophorus, Ageratum conyzoides* etc. Are a threat to native species being more tolerant to climatic variations. 9% of all tree species (woody trees, white spruce) are at risk of extinction. The rich and diversified flora of India provides a valuable store house of medicinal plants. Today, these medicinal plant biodiversity is being depleted due to man made and natural calamities. Climate change seems to be a major driver of decline in biodiversity and particularly loss of medicinal plant diversity and it is predicted to have an even greater negative impact in the decades to come. These climate changes affect medicinal plants variously like phenology- time of plant bud formation and flowering, initial leafing, fruiting, seed setting time, fruit dispersal etc.; secondary metabolite production, quality and productivity due to high carbon dioxide, adaptation etc.

Sudden and rapid climate changes have affected the distribution pattern of animal population worldwide pushing many animals to become endangered and while others to end in extinction. The Polar Regions are already affected by rising temperatures. Diminishing ice packs have reduced the habitats of polar bears, penguins, puffins and other Arctic creatures. Changes in temperature, lower salinities and lower pH have led to phonological, bio-geographical, physiological and species abundance change in marine ecosystems. Corals are also adversely affected by warming of surface water resulting in bleaching to occur and decline in calcification due to increased carbon dioxide. Climate change has serious negative consequences for bird populations. In a recent study, Professor David Bravo- Nogues has said that species have been able to survive new conditions in their habitat by changing either their behaviour or body shape. However, the current magnitude and unseen speed of change in nature may push species beyond their ability to adapt. Until now, scientists thought that species show their reaction to climate change by movement. But, new study shows that local adaptation to new conditions has played major role in the way species survived. Species adapt when the whole population change. For example- when all owls get darker body colour. This occurs very slowly over a long period of time.

## Global Warming or Another Ice Age?

There are some scientists in the world which reject the concept of global warming and climate change. They have a view that global temperature of the earth has not increased since 1995 instead it has decreased since 2002 despite an increase of 8% in atmospheric carbon dioxide since 1995. At the dusk of the 20th century, average global temperature increased at the rate of 1-2$^0$C / century which is under limits of climate change for last 10,000 years. Land based evidences such as moraines and related glacial landforms provided the initial basis for reorganizing the existence of a former 'ice age'. The later discovery of buried soils and deposits containing a fauna indicative of a warm climate beneath or between glacial tills and outwash gravel's, showed that Pleistocene ( about a million years ago) climate has altered between cold glacial and warmer interglacial. In most mid- and high latitude regions ( such as Kashmir, Central Asia and Siberia), conditions during interglacial periods were similar to those of today, although at their peak, temperature may have been up to 2$^0$C above what they are now. Some of the most recent climate change events have been identified as among the potential analogue that have been in the Early Medieval Warm Epoch, the early to middle Holocene, the last interglacial, the late Pliocene and various earlier climates as far as back as the Cretaceous. Temperature is believed to have been relatively stable over one or two thousand years before 1850, with possibly regional fluctuations such as the medieval warm Period or The Little Ice Age. Of all the periods of past climate change, the last glacial -interglacial transition seems most relevant to understand the speed and nature of global warming. According to Milliankovitch theory, changes in insulation (incident solar radiation) provide the driving force for global glacial cycles of melting freezing. The Little Ice Age can be shown to represent a decline in solar irradiant, or a reduction in net radioactive forcing at top the troposphere.

A 2$^0$C change would be well within previous natural bounds. Ecosystems have been adapting to such changes since time immemorial and the result is the process that we term as 'evolution'. The melting of ice in Greenland and Antarctica, the scientists have a view that ice sheets in Greenland and Antarctica are growing in thickness and cooling at their summit. Temperature in the Arctic region is just now reaching the levels of natural warmth experienced during the early 1940's, and the region was warmer during earlier times. In the same way, scientists explain the change in sea level and increase in cyclones and their intensity. According to them, a global average sea level rise of 1-2 mm/year occurred naturally over the last 150 years. All these views seem to lead the view that we are forwarding toward another ice age or the climate is stable. The summer sunlight around the poles is reduced, and high latitude regions such as Alaska, Northern Canada and Siberia become cold enough to preserve snow round the year. This snow  covers reflect back most of the sunlight falling on earth, consequently cooling the earth more rapidly. This is the beginning of new ice age. We now don't know how destructive this ice age will be. Whether this human induced global warming will ensure the easy going life of human beings during the ice age?

The range of atmospheric carbon dioxide in trapped air of glacial ice was 200-300 parts per million (ppm). During this phase, global temperature had fluctuation range of about 12 $^0$C. At present time, atmospheric carbon dioxide level has already touched 400 ppm. This may also increase due to burning of fossil fuels. If the growth rate could be moderated to maintain the $CO_2$ level around 550 ppm, average temperature of atmosphere might increase by 5$^0$C with catastrophic effects on earth like sea level rise and climate change. But, this global warming will not be able to return of ice age as ice age last for thousand of years while fossil fuel might last only for next 300 years. With the passage of time, atmospheric carbon dioxide will decline by getting dissolved into oceans or may convert into carbonate. In about 2000 years, Milliankovitch cycles may induce polar cooling start to coincide again, the present warming trend will has no meaning.

Including uncertainties in future greenhouse gas concentrations and climate change, Intergovernmental Panel On Climate Change( IPCC) anticipates a warming of 1.1$^0$C to 6.4$^0$C between 1990-2100. The countries like China, USA and India have pressure to cut down their carbon dioxide emission and adopt some eco-friendly sources of energy. So, that our earth could remain safe from devastating effect.

## Chapter-9

# Radiation Pollution: Threat to Environment

People's greed to exploit the nature's gifted resources and to become 'super powers' through innovative technologies and to sustain their lives on this earth is increasing the problem of pollution at an alarming rate. Besides, the problem of already prevailing pollutions i.e. air, water, soil and noise, the world is under the exposure of radiation pollution. Radiation pollution is a physical type of environmental pollution. There are many forms of radiation. Some forms of radiation are found in the natural environment and others are due to modern technology. Whether natural or man-made, radiation can be both harmful and beneficial to the environment. The sun, for example can have positive and negative effects on plant and animal life. At low levels, radiation can be beneficial to the environment. On the other hand ionized radiation such as X-rays, gamma rays, alpha and beta particles can be particularly harmful in excessive amounts. The ionizing and non- ionizing radiations released from man made appliances used to enhance the standard of living, contribute to the natural radiation pollution. We all use a fair quantity of electrical goods each day – indeed. It is hard to see how we could ever do without them. Although they make our lives more comfortable and entertaining, they cause huge amount of damage to environment. We all are seduced into wanting the latest model of this or that (electrical gadgets) and we consume and then chunk away vast quantity every day. Just by bringing them into our homes in the first place, we are adding to the problems of pollution in badly. Huge resources of energy and raw materials go into the manufacture of electrical goods, thereby, contributing to climate change. Their disposal and reclamation pose an increasing challenge.

Electrical items produce an invisible vibrating Electro-magnetic field (EMF). Electro- magnetic Field is a combination of electric and magnetic fields that occur in our environment both naturally and artificially. At present the very fabric of our life is being invisibly shredded by waves made by the devices we use. Every person, plant and animal is living under a blanket of man made electro-magnetic smog. Some people consider this field to be harmful to health. It's not just humans who are affected. Honeybees are also vanishing. It is called 'colony collapse' and it has been linked to proximity to microwave towers. Navigational system disorienting and preventing them from finding the way back to their hives or it may be killing in a more direct fashion. Radio waves from cell phones are also lethal to honeybees. We humans can't live without honeybees because they play a critical role in world's food chain pollinating 5% of our food- nuts, seeds, grains and fruits. Thus, the extinction of honeybees would precipitate a global food crisis of almost unthinkable proportion. Ministry of Environment and Forests reported in a study that scavenger birds in certain parts of Orissa have badly affected. Radiations from cell phone towers have a deleterious effect on the avian population. They affect sparrows in the cities and obstruct migratory birds in their flight. These Cell towers harm migratory birds' natural sense of direction affecting their arrival with drop in number. Millions of migratory birds die each year from collisions with telecommunication masts. Number of bats near mobile towers has gone down due to radiation effect. And it is not just bees and birds which are affected, we as well do.

Levels of electromagnetic (EM) radiations have increased more than a thousand fold in last decade especially from mobile telecommunications, electric train tracts, television transmitters and FM radio stations. Electro magnetic field creates contaminants (radio-active particles) that penetrate our air, food, the blood stream and our bones. There is deadly EM smog everywhere. Technology is spewing out electro-magnetism on a scale that increases by the day and it is forming a 'soup' that has many calculated effects on the population. This man made EM smog affects human health. This form of unseen pollution probably causes more harm to humans than all other forms of pollution put together. The dangers of EM radiations are created artificially not only from TV broadcast, FM radio stations, radar satellites and air craft landing systems but also from power lines underground cables, house main circuits, electric trains and motors and household electric appliances, microwave ovens, CFL, cell phones, PDF lap tops, computers, fax machines, CD players and other digital machines, answering machines and even battery operated watches, electric hair dryers, electrical razors etc. or every thing that convert Alternating current (A.C.) to Direct current. Radiation pollution picture is continuously expanding as a complex matrix covering whole earth and we all. These polluting high frequency radiations are detrimental to human health.

The revolution in telecommunication (ICT) in the last decade has introduced the growing use of mobile phones increasing EM radiation pollution. We all are exposed to the mobile tower radiations as mobiles have become an integral part of our lives. Now, we feel that we cannot survive without mobiles. To make these mobiles operational, we consciously or unconsciously and willingly or unwillingly have some mobile towers in our vicinity. Mobile telecommunication companies have established telecom towers in every corner of India facilitating the good communication range to entire population. This network has put the public open to the continuous (24x7) exposure of EM radiations leading to many diseases and disorders of human health and behaviors.

EM smog is an invisible health hazard spreading out of control worldwide. It is an environmental threat as dangerous as tobacco and asbestos and as serious as toxic wastes. As the effects of radiation pollutions are not apparent like other pollutions so it is often termed as silent killer. World Health Organization has said that problem will increase as the technology 'advances', but it is not advancing. How can technology that is causing so much physical, mental and emotional disease be described as being an advance?

## Consequences of Radiation Exposure

Exposure to EM radiations causes extreme effects on human beings as well as on plants and animals. It can kill plants, animals and human beings too or at least cause serious health problems and diseases. Adverse health effects of ionizing Electro- magnetic radiations including somatic and genetic mutations, have been characterized as a result of several studies over the last 60 years. The most disastrous effects are seen with ionizing radiations which are released from disintegration of certain radio-active materials, when they irradiate living tissue, cells are damaged or destroyed. The ultimate effect may be burns, dermatitis, cancer induction, blood change and chromosome damage. Reproduction may also be affected as chromosomes are damaged by the ionizing radiations. This genetic damage could be passed on to future generations as well. Biological effects of ionizing radiations may be in terms of deposited energy in tissues and organs. It may aggravate damage. At Hiroshima and Nagasaki, almost all causalities were due to the blast and fire, but 15-20% was from gamma and neutron radiation. Whole body exposure to large doses of radiation caused severe and  fatal illness. The survivors were liable to die some years later from fatal anemia or leukemia. Other reaction included skin burn, sterility, cataract etc. However, health concerns resulting from exposure to non-ionizing electro-magnetic radiations like Extremely low frequency (ELF), Electric Fields (EF) and Radio frequency (RF) are less well defined. Adverse health effects due to Extremely low frequency and Electro magnetic fields were only shocks and burns resulting from human contact with electric conductors.

Adverse health effects due to high frequency EMR such as  microwave were generally attributed to radiation energy absorption in tissues resulting in excessive heating and consequently thermal damage. Higher magnetic fields are associated with greater risk of leukemia in children and breast cancer in females, increase in brain tumors, miscarriage etc. An intense magnetic field on specific brain regions has been shown to affect thoughts, perceptions and behavior. The most disturbing part of these radiations is their integrating nature. Once exposed to certain radiation level, further exposure adds up to the already received exposure.

There are more recent concerns that cell phone users are more likely to suffer from brain tumors, leukemia, breast, bone and lung cancers, cardiac arrest and even impotency. Extensive use of mobile and the installation of mobile base station in the city and telecom-towers in every corners of the country have put human beings open to the continuous exposure of Radian Exposure War (REW) which is leading to permanent effects on human health and behaviour. According to Bhardwaj and sharma (2011), by maintaining the radiation level up to 450 watt/ square, by installing the base station at a distance of 3 km. away from nearby buildings and by strictly following the basic limit of exposure i.e. "Specific Absorption Ratio (SAR)" the Radiation Exposure War can be avoided. Even exposure to weak electromagnetic fields can disturb the production of hormone melatonin by the pineal glands in the brain, thereby leading to risk of breast cancer and degenerative diseases (like Parkinson, Alzimer, coronary artery diseases etc.). Talking long hours on mobiles may intensify the risk of brain cancer and tumour. Increased genetic mutation in future generations triggered by oxidative stress causing DNA damage is observed in frequent cell phone users. Exposure to Police Officers to high frequency microwave radiations emanating from traffic speed detecting radar units has been associated with increase in testicular, brain and eye cancer. Parietal lobe brain tumors were reported to occur in brain region closest to location of the hand hold cellular phone antenna. Biological facts of EMF include an increase in rate of cell division. It has been found that in case of plants growing close to communication antenna have grown unusually faster suggesting that microwaves were spreading up cell division.

Many scientists have come forward with studies linking EMF to human health risks. People exposed to EMF during their working hours feel changes in their way of thinking and focus having foggy heads increasing headache, emotional changes, frustration and anger, difficulty in learning and remembering, loss of hair, pain in muscles, breathing difficulties, decrease in lactation in lactating mothers, speeding up aging process, decline in immune system etc. Soviet investigators claim that Microwave radiations produce a variety of effects on central nervous system (CNS) and with a temperature rise. Other reported effects include disturbances of menstrual cycle, biochemical changes and changes in RNA. Radiation causes molecules to lose electrons thus destroying it. Killing certain enzymes in the body can simply make us sick. However, once radiation damages DNA, the body may not be able to repair itself. This can increase the chances of developing cancer both in animals and humans.

Electro-smog is known to cause acidification of body through manufacture of free radicals. Exercise gym can subject visitors to a dense field of electro-magnetic pollution through all the electronic exercise machines. Dr. Macada Havas from Trent University, Canada published the relationship between EMF and diabetes. Most interesting findings in her study was that electro sensitive people whose blood sugar decreases when they go for a walk outdoors actually experience an increase in blood sugar when walking on treadmill. Radiations also affect the foods we eat. EM radiations cause chemical changes in foods resulting in free radicals such as benzene, formaldehyde and hydrogen per oxide. These changes may be considered as the potential cause of cancer.

Children are more at risk than adults because of easy absorption of microwaves. Because their immune system is developing, kids have more absorption. Genetic damage is more in children due to increased cell division. So, children under sixteen should not use cell phones. Investigators at John Hopkins University have suggested a possible relation between magnolism (Down's syndrome) in offspring and previous exposure to male parent to radar. Radio frequency radiations badly affect the foetus. When the pregnant lady either uses mobile phone or when illuminated with radio frequency radiations, the developing embryo may have developmental malformation. It also produces thermal heat which changes the spread of biochemical reactions possibly the functions. Due to over exposure to EM radiations, several symptoms of various respiratory, cardiac, ophthalmological, dermatological and neurological disorders occur. Hiroshima, Nagasaki, and Chernobyl, all these three places were exposed to a tremendous amount of radiation and the people living here suffered the devastating effect of radiation which up to this day is still being experienced by the victims or their offsprings.

Radiation pollution has bad effect on both land and to human health. The land that was exposed to radiation will be unfertile. Soil can become compacted and lose the nutrients needed for plants to grow. It would not be fitted for farming anymore and probably no living organisms would survive. As exposure increases, plants become sick and stunted and will die. High levels of exposure will kill trees. For example- conifers are more sensitive than deciduous trees. Natural radiation is often beneficial to plant growth. It is necessary for many plants to receive some form of non-ionizing radiation. Radiation that produces light in order for photosynthesis to occur is a positive effect that radiation has on plant life. However, according to the Environmental Literacy Council, ionized radiation that occurs from nuclear material may result in weakening of seeds and frequent mutations. For example, the leakage in Chernobyl nuclear plant in Russia in 1986, caused excessive amounts of radiation pollution in that region. A huge cloud of radiation was formed which resulted in a massive amount of destroyed plant life; particularly pine trees in that area. High doses of radiation can be devastating to the environment. High levels of ultraviolet radiation (UV) can cause a reduction in reproduction capabilities in marine animals. It can also disrupt the timing in that plants flower, which can result in changes in pollination patterns. According to NASA, it can also reduce the amount of food and oxygen that plankton produces. Plankton can respond to excessive amount of Ultraviolet-B (UV-B) light by sinking deeper into the water. This decreases the amount of visible light required for photosynthesis, which reduces growth and reproduction. An increased amount of UV-B can also increase the amount of ozone produced at the lower atmosphere. While some plants can use this extra layer as a protective shield, while other plants are highly sensitive to photochemical smog.

Radiation pollution is very dangerous not just to our nature but also to us. As the exposure to radioactive materials is severe and bears grave consequences, so radiation pollution prevention is mandatory. The activism about EM pollution should not be seen as a war between environmentalists and companies or health organizers but as a chance to improve the quality of life by using new more well designed and sophisticated technological products. It is a chance to rebuild the existing market on a healthier basis with realistic prospects for everyone.

NASA studies have proved that household plants are efficient filters in indoor world. Some plants may offer some protection against electro- magnetic radiations and help in retaining comfortable moisture levels at home. The plants not only look beautiful but help to lessen the feeling of detachment of nature that affect so many of us in modern society. Natural ecofriendly and antipollutant house plants include – Peace Lily, African Violets, Christmas cactus, Garlic Vine and various Ivies.

People have dubbed this 'EM' Smog like real smog; it can have serious effects on our health. So, avoid spending long hours near man made gadgets emitting EM radiations. By thinking more about the way we buy, use and dispose the electrical goods, we can't only cut down pollution inside home but improve our health and that of the environment as well. No doubt, nuclear energy has proved a boon for cheap, eco-friendly and large power availability but it may become a bane with a small shift in nature. Therefore, radiation effects, hazards and safety concerns must be considered while using radiation energy for peaceful alternate endeavors and remedial measures should be taken by all of us to ensure clean, green and electromagnetic radiations safe environment.  Now, it is our duty to see that our life support system and environment has not degraded by our activities and we have to hand over a healthy and clean environment to our coming generations. If we have to survive, each human being should contribute in creating positive environment.

# Chapter-10

# Polythene Pollution

Today, polythene/plastics are used so universally that they have become a major part of our everyday lives. Use of polythene is increasing day by day and now it can be seen that polythene is being used in almost every activity of life. We have engulfed with plastic products and materials which have become the most attractive options for varied applications in every field of our life. It is realized that 'plastic revolution' has been going on in our lives or we can say that we are living in 'plastic era'. Plastic/polythene has become a human necessity and life appears to be inseparable from world of this man made material -the plastic, because of its immense benefits. We can hardly do away with it. Everyday, we come across plastic in various forms from the tooth brush we use daily in the morning to plastic bucket, plastic chairs, cups etc. to telephone/mobile to chat, to televisions and computer/ lap tops. Plastic is rapidly replacing wood, glass, iron, ceramic crockery and even construction materials in industry and household items. Today it is used to make or wrap around many of items we buy or use. Modern life is entirely based on plastics be it cell phone, televisions, syringes, wire insulations, laptops, car dashboards etc. The problem arises when we dispose off them. It is very cheap and gets discarded easily and its long life provides survival in the environment for long time where it does harm. It is found anywhere we go, it piles up in streets of cities, in drains, floating on the water bodies and so on. The list is unending! Polythene is a necessary evil. Besides humans, animals were discovered using plastic wastes as a building material for their nests in urban areas. Indian palm Squirrels (Funanbulus palmarum) are using plastic bags, cigar buts along with natural materials in urban areas as nesting materials instead of leaves, twigs, bark, mosses and other soft materials. With the increased use of polythene, the pollution level caused by it will also be increased, affecting almost every kind of environment including terrestrial as well as aquatic biomes. In urban areas, polythene is a major threat to both environment and public hygiene. The problem is so acute that it has become a priority issue in maintaining better management and protection of ecosystems.

Polythene is typically an organic polymer of ethylene gas $CH_2=CH_2$ which is commonly used in our daily to day life like grocery bags, shampoo/cream bottles, cables, various domestic items etc. Polythene is unaffected by any acid, base and other organic solvents. Many molecules of ethylene are joined to each other in presence of oxygen catalyst at high temperature and high pressure.

$$nCH_2=CH_2 \xrightarrow[\substack{100-300^\circ C \\ 1000-3000 \text{ atm.}}]{O_2} -(CH_2-CH_2-)n-$$

The word, 'Plastic' is a household term that has came from Greek word *'plasticos'* or Latin word ' Plasticus' which means- to be able to moulded or shaped by heat. 'Polymers' is a generic term for all plastic material referring to organic carbon based compound. India has a large share of 4% of total plastic consumed worldwide. Versatile nature of low density polythene makes it major cause of pollution. Polythene/plastic pollution is an example of non-biodegradable pollution and they persist on our earth years after years affecting even future generations. We have invited ourselves to a plastic party with a devastating outcome. It is an epidemic now. It has drastically disturbed every man's life style.

## Environmental Toll of Plastic/Polythene

Although the application of plastic has facilitated in preserving our natural resources (Wood), conservation of energy etc., plastic bags and films used in nursery bags or in green house have increased yield in agriculture field, plastic waste is an environmental problem in today's society. The rapid increase in population and urbanization in India has led to an increase in its use as well as waste generation. In most of the cities, a large quantity of plastic finds its way to landfill while in rural areas, it can be seen scattered among vegetation. Plastic does not decompose easily, persists in environment, and pollutes the land, water and air. It not only harms terrestrial flora and fauna, it has been known to have a disastrous effect on aquatic environment as well. Not only oceans are contaminated but seabed near the coastal areas is also found to be rich in plastic bag contamination. It poses a serious threat to entire population of many marine species as well as health risk to humans also. The threat to marine environment is caused by marine debris of plastics. Plastic pollution in marine environment is the major cause of several hazardous and ecologically damaging effects. Everyday, more and more plastics is accumulating in our oceans from two sources- land based sources and ocean based sources.

In Land based sources, plastic enter the marine environment from sewage outfalls, merchant shipping operations, tourists and beachgoers. The majority of the shoreline debris is now made up of plastic waste. A greater concentration of plastic material is found near popular tourist destinations and densely populated areas. Actually ocean pollution starts out on land is carried away by wind and rain to the sea. Plastic wastes can easily be transported to long distance because of their low density. According to a study, around 10-20 million metric tons of plastic ends up in our oceans every year. Scientists have warned that there will be more plastic than fish in the ocean by 2050 unless a serious effort is made to reduce our dependence on plastic items. Ocean based sources include garbage dumped at sea by ocean going vessels and fishing debris like fishing line, fishing nets etc. Abandoned fishing nets drift about endlessly in the ocean –indiscriminately trapping marine organisms that swim into them. Discarded fishing line can entangle birds, turtles and other marine life while larger marine animals such as seals are vulnerable to being 'noosed' by bait box strapping.

People recklessly throw plastic bags, straws, paper cups etc. in water bodies or animal enclosures without realizing where it will end up. As a result, forest areas and water bodies became dangerous for wildlife. Plastic can affect marine wildlife in two ways- first by entangling creatures and second by being eaten. Plastic bags look like a tasty jellyfish to an indiscriminate feeder like sea turtle and eat them but plastic is indigestible. It can choke or block the intestines or cause infection in those animals that consume it. Turtles have found dead with plastic bags in their stomach. South African sea birds are among the worst affected in the world. Plastic may remain in stomach blocking their gastric enzyme secretion and diminish feeding stimulus ultimately causing starvation, injury to internal organs and finally reproductive failure. A Bryde's whale died in August 2000 on Australian coast. The autopsy revealed that whale's stomach was tightly packed with plastic like supermarket bags and food packaging materials etc. About 267 species globally including invertebrates, 86% includes sea turtle, 44% fish, seabird species and 43% marine mammal species have been reported to ingest or become entangled in plastic debris resulting in impaired movement, reduced feeding, injury to internal organs, reduced or failed reproduction and ultimately death. As per a study conducted in USA, out of 1033 birds collected off the coast, 55% of them had plastic in their digestive system. This is because of their habit of selecting their prey on the basis of colour, therefore,  polythene is mistaken for prey. Being an important food item for many species of mollusks and crustaceans, plastic ingested by invertebrates may transfer the toxic substances up in food chain. Besides, while diving for food, both seal and whale get caught in translucent net and drown. Northern fur seal often play with fragments of plastic netting or packing straps, catching their necks in the net. This plastic harness may constrict the seal's movement ultimately killing the seal through starvation or infection due to wounds. Royal terns (Sterna maxima) are among the species of sea birds that dive from air to water to catch fish with their sharp beak. A plastic bag floating on water surface become invisible to tern and may have attracted fish to the first place. Plastic debris also provides favorable condition for potentially invasive and destructive fauna to migrate to new environment. Colonization and subsequent dispersal of marine species is common with many barnacles, bryozoans, polycheats, tubeworms, dinoflagellates, algae and mollusks found to adhere to plastic debris. Reports are also available for terrestrial fauna, such as Ants - reported to ride the debris from Brazilian mainland to San Sebastian Island, several kilometers away. Persistent organic pollutants (POPs) such as dichlorodipheyltrichloroethane (DDT) and polychlorinated biphenyls (PCBs) - both harmful endocrine disrupters, can accumulate on the surface of plastics that are million times higher than concentrations naturally occurring in seawater. These pollutants are stored in the fatty tissues and organs of the body and get passed on up the food chain, becoming more concentrated in animals at the top of the marine food webs. Therefore, ocean's top predators such as polar bears and killer whales become more contaminated organisms suffering from reduced fertility, lower breeding success, weak immunity etc. Dr. Lamb from Cornell University in USA has highlighted the impact of plastic pollution on marine ecosystem especially on the establishment of the delicate coral and coral reefs. The presence of these non-degradable plastic wastes in the benthic floor enhances diseases among keystone reef building corals in the Asia –Pacific region. More and more plastic debris is entering the ocean every year. This debris in turn become overloaded with pathogenic bacteria and causes a disease among coral known as "White Syndrome". Corals are also suffering from more fatal diseases like 'Skeletal eroding band disease' and 'Black band disease' due to constant contact of plastics with tissues of corals. Besides, plastic harms the corals by ushering in shading or low light micro – environment leading to anoxic conditions favouring diseases, consequently declining coral reefs. There are various case reports of death of cows in rural as well as urban areas due to consumption of polythene bags. This remains undigested in their gut, thereby leading to blockage of digestive tract. In India, about 1,00,000 animals such as cows, dogs, penguins are killed every year due to eating polythene bags.

Those plastics which are degraded either mechanically or chemically are ultimately reduced to tiny pieces of the size of grain. These tiny particles or micro plastics (particles smaller than 1.5 mm. in diameter) spread over long distances through ocean surface circulation and then ingested by various small creatures. These plastic particles may persist in  aquatic environment for decades/ centuries and may concentrate persistent organic pollutants (POPs) present in seas. Plastic is photodegradable- it becomes brittle and breaks down into smaller particles when it is exposed to Ultra Violet radiations from sun. According to UN Environment Programme Executive Director Ached Steiner, 'Marine debris- trash in our oceans- is a symptom of our thrown away society and our approach to how we use our natural resources.'

## Micro Plastic

Micro plastics have now become a rapidly increasing problem. Polyethylene (PE), Polypropylene (PP), Polyethylene Terephthalate (PET), Polymethylmethacrylate (PMMA) and Nylon (PA) were found to be most common micro plastics. 'Primary micro plastics' are found in personal care products such as shampoos, toothpastes, gels, pillows (polystyrene), synthetic clothes (polyester, nylon) and vehicle tyres. Heat and pressure cause tear of tyres producing micro plastic dusts. Washing of synthetic clothes releases micro plastic fibers in the waste water. Decorative glitters used by school going children have generally PET and PVC. Tea bags also contain polypropylene fibers. The outer fuzzy layer of tennis ball is made up of PET micro plastic. Paper cups have a lining of polyethylene. The paper decomposes leaving behind tiny micro plastics. Breakdown of macro plastics due to Ultra Violet radiations generate 'Secondary micro plastics', large amount of which is persistent in water bodies. Micro plastics also absorb other organic pollutants present in water bodies and act as sponges. Therefore, they act as vectors, transferring chemicals from the surrounding environment into marine food chains.

Among all the metro cities of India, Delhi is on the top for generating plastic wastes.  Most of the plastic bags go into landfill. They find their way into water ways, parks and streets. If burnt, infuse air with toxic fumes causing air pollution with breathing problems. Degradation rate of polythene is very slow in landfills due to anaerobic environment. Land filling has its own associated drawbacks as land remains unavailable for a very long time. Otherwise the land may be utilised for agriculture and other related activities. Polythene debris remains for very long time in landfill and it has been published in a report that it will take around 500 years to breakdown in landfills and leaving the land barren. This secondary environmental pollution has become a major problem of landfills. Pollutants are released in the form of leachate as well as gases being toluene, benzene, xylene, ethyl and trimethyl benzene etc. Besides, plastic debris deep in landfill can leach harmful chemicals that spread into groundwater.

As polythene bags do not decompose with soil, they remain in agriculture land and block as well as retard the growth of agriculture plants. Very thin roots of crop plants fail to pierce them in order to move around the soil for natural nutrients. It degrades the land quality. Thus, polythene bags have caused a great harm to agriculture plants posing threat to our food and life. The various impacts of polythene bags on agriculture are- reduced soil fertility, reduction in yield, reduction in nitrogen fixation, loss of nutrients in soil, disparity in flora and fauna on soil etc. all these negative impacts ultimately lead to reduced agriculture production. When large amount of plastics are mixed with soil, they affect the growth of earthworms in soil and reduces soil fertility. When plastic bags stick to trees and plants, they obstruct the natural light from reaching different parts of the plant leading to impaired photosynthesis. Burring of plastic bags in soil makes an insulating layer that keeps rainwater in upper part of soil and prevents water and other fertilizers from reaching its lower part.

## Effects of Polythene Pollution on Human Beings

The earth in which we live and grow our food, air we breathe and water we drink has an effect on our health. A study of US Centre for Disease Control Prevention opined that 93%of US population contains bisphenol A, a chemical that can be found in canned goods and hard plastic items in their body. Bisphenol A is key component in production of polycarbonate plastic which is used to make a variety of products- water and baby bottles, sports items, dental devices, dental fillings and lenses etc. Bisphenol A, an endocrine disruptor, is when absorbed by human beings and wildlife mimics the activity of hormones. These chemicals have been associated with reproductive problems in human as well as in animals. Exposure to these toxic bisphenol and phthalates can be associated with increased risk of diabetes, pollution-related chronic diseases, cardiovascular disease, cancer, and hypertension in human beings.

Biomagnification, or bioaccumulation, is a natural process and can be described as the increasing amount of a substance, such as a toxic chemical, in the tissues of organisms at successively higher levels in a food chain. In nature, this process is supposed to be beneficial to organisms higher up on the food chain, like major predators. However, because of the presence of toxins such as polychlorinated biphenyls (PCB's), nonylphenol (NP), dichlorodipheyltrichloroethane (DDT), polycyclic aromatic hydrocarbons and Bisphenol A(BPA) etc. and waste in the ocean, certain chemicals can bioaccumulate in organisms upon consumption and travel up the food chain, its concentration increases and finally eaten by human. Because many people eat fish, which have bioaccumulated chemicals in their meat, amounts of these chemicals transfer into human bodies. Seafood, including shellfish, fish and other marine organisms, is an important source of protein for humans across the world. Direct toxicity from plastics comes from lead, cadmium, and mercury. Diethylhexyl phthalate (DEHP) contained in some plastics, is a toxic carcinogen. Other toxins in plastics are directly linked to different types of cancers (Prostrate, ovarian and breast), birth defects, hormonal imbalances, immune system problems, and childhood developmental issues, endocrine disruption, DNA hypomethylation and neuro- behavioral changes etc. Polythene can also cause AIDS and thrombosis. Ultimately, human health is threatened by the bioaccumulation of toxin because the chemicals ingested can cause cancer, developmental issues, or other problems.

Use of plastic containers to save food items particularly pickles, fatty foods causes plastic to decompose and toxic carcinogens to access the body. Nowadays, most popular polythene pollution is caused by polyvinyl chloride (PVC). When any food material is stored in it, then gradually the soluble chemical gas dissolved in them cause death due to cancer and other skin diseases. PVC has been found to destroy the fertilization of animals and human beings and their reproductive systems.

## Possible Effects of Plastic Debris in the Future

Plastic is a wonder material having endless uses, but a problem arises everywhere because after it has served its purpose, we throw this away. Plastic waste is increasing due to increase in population, developmental activities, changes in life style and socio- economic conditions.

The abundance of plastic in the environment will have deleterious ecological effects that become more prominent in the future. One of these effects may be the rapid spread of diseases. Both land and aquatic plastic waste has the potential to carry hosts from non native lands and threaten biodiversity. Plastic can serve as a transportation device for pathogens, bacteria, or organisms from foreign lands. This spread of invasive species can threaten the stability of ecosystems.

The bioaccumulation of micro-plastic is becoming an increasingly important for seafood industries and the general population because it poses a toxic health hazard for every organism involved. Once the micro-plastic accumulates in an organism and it is consumed by another organism on a higher trophic level, these particles then get absorbed into the consumer's body.

## Polythene Disposal

There is an urgent need to find out the proper method for polythene disposal. Currently four considerable options are available viz. thermal treatment, land filling, recycling and biodegradation.

### (a) Thermal Treatment

Thermal decomposition of polyethylene results in evolution of harmful gases during fires or waste burning like carbon monoxide, chlorine, furans, dioxins, CCl4, etc. resulting into breathing problems. Therefore it is mainly done through incineration and pyrolysis. Plasma Pyrolysis is an effective method to destroy polythenes in an eco-friendly manner.

## (b) Landfilling

Polythene debris remains for very long time in landfill and leaving the land barren. Pollutants are released in the form of leachate as well as gases viz. toluene, benzene , xylene ethyl and trimethyl benzene  are very much harmful for human health.

## (c) Recycling

Recycling of plastics involve many processes including mechanical, chemical and thermal depolymerisation. It has been concluded by several researchers that during recycling more toxic and larger amounts of hazardous volatile organic compounds are emitted from melted virgin and waste plastic pellets than producing virgin plastics. But recycled plastics have proved to be more hazardous than virgin plastic as during recycling it is mixed with a number of harmful colours, additives, stabilizers etc.

## (d) Biodegradation

The focus of people has shifted to biodegrade the plastic with the help of various microorganisms and recovers value from polythene. In biodegradation, strong carbon bonds are broken down through microbial actions that reduce the strength of polythene (as molecular weight decreases) and hence polythene gets degraded.

## What Can We Do?

The plastic pollution is one of this century's major issues impacting the environment health of our planet. Each year, 500 billion to 1 trillion plastic bags are consumed worldwide. Banning Plastic bags has been a key step in moving towards a sustainable society globally. There should be strict rules for limiting the use of polythene to a certain amount. In India, several states and municipalities have imposed a total ban on the use of plastic bags, straws, plastic beverage bottles and other disposable items while others have implemented control measures to curb their use.  It should be replaced by paper or cloth bags, glass bottles and cups etc.  We consumers can play a vital role here. We have to refuse the disposable food packaging instead we can choose reusable containers. We should take every opportunity to reduce our use of plastics. The Ministry of Environment, Forest and Climate Change has banned plastics in all protected areas around the country and has declared them as "plastic free zones". A ban on polythene bags below 50 microns in thickness was implemented in 2016 across the country. A digital campaign "Green Good Deeds" has also been launched with the aim to encourage people to take anti-pollution measures such as planting a sapling and recycling waste. India has introduced new plastic waste management regulation in March 2016 in an effort to reduce the amount of plastic waste its citizens generate.

Awareness camps should be organized in every corner of the city to aware people on restricting use of polythene and harmful effects of it and to use biodegradable reusable bags only. Polythene waste that is non biodegradable can be used for other purposes like preparation of bitumen after mixing with asphalt that can be used for road construction. Jamshedpur Utilities and Services Company Limited working for Tata Steel, has started to use plastic debris for road construction in Jamshedpur, Jharkhand. 10% of plastic is mixed with coal tar. This plastic works as binder delaying the cracks in the road.  A student of a school in village Garudvasa of Jamshedpur Telco has made an eco-friendly toilet using 11000 waste plastic bottles. Similarly, an organization- Dharti Maa Trust in Faridabad, Haryana has started making earten glasses ( Kulhar) and plates from Leaves of Dak ( Butea monosperma) as an alternative to thermocol and plastic glasses and plates. Researchers of Britain and America have invented an enzyme named as ' PET-PAGE' by which Pet bottles can be recycled and production of plastic may be reduced. MARPOL is an important legislation restricting discharge of garbage  and bans in sea disposal of plastics and other synthetic materials such as rope, nets, bags etc. Legislation at national level also plays an important role. Laws that encourage recycling must be enforced. Banning of food packaging on beaches can also check this problem. The most important is to bring about behavioral changes and awareness to stop the flow of plastic waste into aquatic habitats.

Researchers have been focusing their efforts to find ways for bio-remediation in the environment. In this regard, microbes are pivotal as they produce enzymes capable of degrading plastic polymers. One such microbe, Ideonella sakaiensis- a bacterium – identified from soils of plastic recycling plants in Japan seemed very promising. It is found that this bacteria may eat soda bottles made up of Poly- Ethylene- Terephthalate (PET). This bacteria was found to have adapted to produce an enzyme that is able to partially degrade the plastic polymer to individual monomers and use that as a source of energy. PET degrading enzyme PETase produced by this bacterium breaks down highly sturdy PET polymer. Similarly, a fungus, *Aspergillus tubingenesis* has been discovered by scientists of the Chinese Academy of Sciences (CAS). The fungus was found in Pakistan and usually lives in soil, but can also grow on the surface of plastics. This plastic eating fungus secretes enzymes onto the surface of plastics breaking the chemical bonds between plastic molecules, or polymers. Scientists are trying to identify the ideal conditions to replicate this process and provide a sustainable solution to get rid of ever increasing non- biodegradable plastic dumps. These findings raise hope that in near future, the plastic hazard in the environment will be resolved.

However, the use of plastics can be reduced in the ways given below-

☆   Always carry cloth shopping bags while going outside.

☆   Avoid keeping hot food substances in plastic containers.

☆   Try to use utensils/ glasswares /earthen wares instead of disposable plastic wares.

☆   Dispose plastic wastes in proper bins.

## Bioplastic

Synthetic plastic are primarily hydrocarbon in nature and are non-biodegradable. The Green Chemistry Centre for Excellence at the University of York is working on synthesizing 'bio-based polyesters' for clothes and packaging. Polylactic acid (PLA) derived from corn, cassava or sugarcane and Polyhydroxyalkanoate made by micro-organisms that produce plastic from organic material are biodegradable. Toyota Prius uses bio-degradable plant derived (sweet potato) plastic for interior car components since 2009. These bio-degradable plastic break down anaerobically in landfill releasing methane. But, it is in initial stage and also cost three to four times the cost of non-biodegradable plastics.

Now, it looks very difficult to replace plastic totally, a continuous effort must be made to reduce its use for conserving our environment. It seems to be harmless but in reality, it is a silent killer. 'Refuse, Reduce, Reuse, Repurpose and Recycle are the only way out. The Earth day, 22nd April, 2018 was dedicated to spreading awareness about the pollution caused by plastic and the need to eventually end it. With " Beat Plastic Pollution" as the theme for World Environment day, 5th June 2018, the world is coming together to combat single use plastic pollution and India was the global host of World Environment Day. Dr. Harsh Vardhan, Minister of Environment, Forest and Climate Change, said," It's a global emergency affecting every aspect of our lives. It's in the water we drink and the food we eat. It's destroying our beaches and oceans. India will now be leading the push to save our oceans and planet." The theme urges governments, industry, communities and individuals to come together and explore sustainable alternatives and urgently replace the production and excessive use of single –use plastic. Efforts should be taken to mitigate this problem. Legislation along with education can prove to be strong step towards coping up with serious threat of our environment. General public and scientific communities together can work in hand to solve the problem. We can do our bit. It starts with each and every one of us. It is now time for us to take immediate action.

"Let us contribute our part to save our environment from polythene pollution and make it a better environment for future."

# Chapter-11

# Toxic Tour in Homes

Pollution is an undesirable change in physical, chemical and biological characters of environment which affects human life. Today problem of pollution has become a major international challenge. But, pollution knows no boundaries. Traditionally, the source of air pollution in cities brings to mind mainly vehicular and industrial pollution. However, recently we don't need to go the open world to be polluted. It is imperative. It destroys our health even inside of our homes which we consider as safest. The source of air pollution in urban spaces has shifted to volatile chemical products and 60% of all chemical pollution is within house. To avoid outdoor exposure of polluted unhealthy atmosphere, we protect ourselves spending most of our time at home. Unfortunately the indoor environment can be quite harmful to our health than the outdoors. Indoor environment is two to five times more toxic than our outdoor environment and on an average, people are spending about 90% of their lives indoor.

Indoor places- our homes and workplaces are closed and have almost airtight environment in which air circulation is within indoors containing air contaminants. These contaminants are invisible but act as silent killers. Our modern day homes are rife with pollutants that can cause a host of health problems. Chemicals from furniture, floorings, paints, detergents, household cleaners can hover inside poorly ventilated homes as well as with good air flow. We inhale this polluted air regularly, it penetrates deep into the lungs and cardiovascular system causing extreme health risks. Though the air contaminants are lesser in volume than the outdoors, they impact severely many times than outdoor contaminants. WHO reports that indoor air pollution is the second largest cause of morality in developing countries like India. The visible indoor air pollutants are biological components like fungus, molds, fur, mites, pollens, fumes, animal faeces in air etc. all these elements contribute to suspended particulate matter(SPM) in indoor environment. Improper ventilation also acts as a catalyst for increasing the levels of SPM. Environment Tobacco smoke (ETS) represents another relevant source of indoor chemical pollution. It is a mixture of smoke that comes from the burning end of cigar, pipe or cigarette and smoke inhaled by the smokers. It is a complex mixture of 4000 compounds, more than 40 of which are carcinogenic to humans and animals. In adults, ETS has been found associated with coronary heart diseases, irritation and exacerbation of respiratory symptoms including asthma. In children, middle ear infection and ETS has been observed.

Another contributor of indoor chemical pollution in urban areas is our life style. Many of us use so many toxic chemicals in everyday life; sometimes we are aware of it, sometimes not. We are exposed to a variety of synthetic volatile organic chemicals emitted from our domestic items like carpets, wall coverings, building materials, dry-cleaned clothes and combustion gases. Besides, everyone likes a clean home but we often rely on hazardous substances to sparkle or shine our floor and bathroom. We lock these domestic utility substances in closets or under the sink to keep them away from children but we don't consider what they may be doing to our health or home. Some of them are not just poisonous but carcinogenic too.

Let us start toxic tour-

## In the Kitchen

Generally, an all purpose cleaner, ammonia based cleaner, bleach , brass or other metal polishes, dishwashing detergents, disinfectant, drain cleaner, etc. are used. These contain dangerous chemicals. Some examples are –

- ☆ Sodium hypochlorite in chlorine bleach: if mixed with ammonia releases toxic chloramine gas. Exposure may cause mild asthematic symptoms or respiratory problems.

- ☆ Ammonia in glass cleaner : may cause headache and eye irritation.

- ☆ Phenol and cresol (in disinfectant): can cause diarrhea, dizziness, fainting etc.

- ☆ Formaldehyde : a preservative in many products. Suspected human carcinogen, strong irritant to eyes, throat, skin and lungs.

## In the Utility Closet

A number of products are likely to contain toxic ingredients- Carpet cleaner, room deodorizers laundry detergents, anticling sheets, mothballs, mosquito repellents, spot removers etc. all not only add aerosols to indoor air but usually contain toxics chemical like –

- ☆ Perchloroethylene or 1-1-1- trichloroethane solvents (in carpet cleaner) can cause liver and kidney damage.

- ☆ Naphthalene (in mothballs) - suspected human carcinogen.

- ☆ HCl or sodium acid sulphate in toilet bowl cleaner - either can burn skin or cause vomiting diarrhoea.

- ☆ Residues from fabric softerners as well as fragrance used in them can be irritating.

## In Living and Bed Rooms

Even furnishing of a typical home is made up of wrinkle resistant polyester blend fabric treated with formaldehyde resin. These also include curtain sheets etc. Modern furniture made up of pressed wood emits formaldehyde and other chemicals. Many carpets emit a chemical called 4- phenylcyclohexane, an additive to latex backing used in commercial and home carpets.

Besides, soft vinyl toys, baby bottles, rattle, teething rings, plastic bottles are made up of polyvinyl chloride (PVC) softened with phthalate plasticizers. These are conduits of cancer causing chemicals. Children are at risk from liver and other organ toxicity. Teflon coated non stick utensils have also a bad reputation for emitting toxins.

## In the Bath

Numerous cosmetics and personnel hygiene products contain hazardous substances like -

☆ Cresol, formaldehyde, glycols, nitrosomine and sulphur compounds in shampoos.

☆ Aerosols propellents , ammonia, alminium chlorohydrate in antiperspirants and deodorizers.

☆ Glycols, phenols, fragrance and colours in lotions, creams ,and moisturizers.

Detergents, stain removers, pesticides and insecticides get carried indoors, adding to further contamination and have made our homes a miniature chemical factories. A study conducted by environmental engineering at Texas University at Austin reports that showers and dishwashers also contribute to indoor air pollution. A new study claims that shampoo, cleaning products and even perfumes could be contributing to harmful air pollution as much as traffic fumes. Chlorine bleach, disinfectants and other household cleaning sprays can spark asthma and worsen existing symptoms. Jan-Paul Zock of the Centre for Research in Environmental Epidemiology in Barcelona has warned that professional cleaners and health workers who use products in hospitals are particularly vulnerable. Previous studies have already found higher rates of asthma among caretakers, cleaners, housekeepers and nurses. Inhaling bleach, ammonia, decalcifiers, acids, solvents and stain removers more than once a week was linked to a 20% rise in asthma or wheezing. Trace amount of chemicals like radon, petrochemicals and by-products of chlorine treated water get transferred to indoor air through volatilization. The study shows that when such water undergoes heating, the released vapours enter into respiratory track. Hence, damage caused to lungs result in chronic bronchitis and non-reversible asthma. Besides, indoor air pollutants may cause skin, eye and throat irritation, nausea, dizziness, headache, nerve and digestive disorders.

There are so many easy ways to detoxify our homes and bodies. Until 1950, the house holders used a limited number of simple substances to keep most objects clean, odour free and pest free. Soap ,vinegar, backing soda, washing soda, borax, corn starch and certain food ingredients were used to lift out spots and stains, deodorize, polish wood or metal disinfect, starch clothes and to perform countless household tasks. Simple cosmetic preparation were made with aid of eggs, oils, clay, herbs etc.

Buildings of past were made with wood, brick, stone, cement etc. Furniture was made with solid wood, carpet with wool or cotton. In many urban buildings, toxic materials like lead and asbestos were used, both known now to be fatal to human health. People experience acute health symptoms while residing in building due to residues left behind by newly constructed buildings which contribute air borne building contaminants. The symptoms of this 'Sick Building Syndrome' (SBS) disappear when the occupants are away from the buildings. The common complaints of SBS are headaches, eye, nose or throat irritation, cough, dry and itchy skin, dizziness, inability to concentrate, fatigue and sensitivity to odour etc.

We have little control on pollution outside. We do not need to return to ways of past to avoid exposure to house toxics but we can take some lessons from past for a better and safe future. We need to consider the ubiquitous use of cleaning products at home. There is need to learn some ways we can prevent chemical pollution for us and our family.

## Alternative to Hazardous Household Chemicals

Toxic chemicals in home can be eliminated simply by making thoughtful choices after educating oneself about where the hazards are in common consumer products. One shelf of simple and safe ingredients can be used to perform most home cleaning tasks. What needed is knowledge of how they work and how different substances should be combined to get the cleaning power needed for particular jobs. Finding non toxic alternatives are always best and there are recipes for them.

### What can we do?

1. *Deodorizer -*     use baking soda (sodium bicarbonate) as deodorizer in refrigerator, on smelly carpets, on upholstery etc.

2. *Copper Cleaner -*     rub with salt and lemon juice mixture.

3. *Silver Polish -*     rub with toothpaste on a soft cloth.

4. *Furniture Polish -*     rub with olive or almond oil.

5. *Ceramic Tiles -*     mix 1/4 cup baking soda, 1/2 cup white vinegar and 1 cup ammonia in gallon warm water (good general purpose cleaner).

6. *Drain Opener-*     use plumber's snake, pour boiling water down drain.

7. *Upholestry Cleaner -*     clean stains with club soda.

8. *Spot Remover -*     for butter, coffee, gravy or chocolate stains, dab with cloth dampened with a solution of one teaspoon white vinegar in one quarter of cold water.

9. *Toilet Cleaner -*     pour ½ cup liquid chlorine bleach .

10. *Moth Repellent -*     use bay or neem leaves.

11. *Ants -*     spread cinnamon, red chili powder, cream of tarter.

# Indoor Plants – Valuable Tool for Removing Toxic Chemicals from Indoor Air

Mother Nature truly is a wonderful woman. There is an easy and affordable way to combat the presence of yucky stuff we may breathe in and it comes from the natural world. Common indoor plants may provide a valuable weapon in rising level of indoor air pollutants. They are not only decorative, but add freshness also and National Aeronautics and Space Administration (NASA) scientists are finding them to be useful in purifying air by absorbing harmful gases and toxic volatile organic chemicals like stubborn formaldehyde, benzene, and carbon mono oxide etc. from indoor air. Bill Walkerton of NASA said 'plants take substances out of air through the tiny openings in their leaves' and NASA calls 'nature's life support system'. The natural photosynthesis breaks down some of the chemicals, further assistance is provided by micro-organisms present in the soil. The phytoremediation is beneficial not just for indoor atmosphere but outdoor as well. The flora require minimal care and maintenance, they not only absorb the toxic chemicals but synthesize them to release oxygen. Beyond air quality, plants just make people feel better. For example, hospital patients with plants in their rooms were more positive and had lower blood pressure and stress levels. Similarly, indoor plants may make people smarter by allowing them to stay alert and reducing mental fatigue.

According to American Ecologist and founder of modern environmental movement, Barry Commoner, Environmental pollution has become an incurable disease and it can only be prevented. As we move away from the nature, problem will be more serious. Also, the economic variations among people lead to unhealthy practices deteriorating the condition further. There is urgent need for government to identify the categories and groups of chemicals that are currently in extensive use in daily utility products that are major contributor of chemical pollution. Efforts must be made to identify affordable and less harmful alternatives. The way for recovery from issue of pollution lies in a collective social responsibility. We all should be aware in making conscious choices. Every individual should contribute to safeguarding the present as well as the future generations from atmospheric pollution.

Some Indoor Plants- Natural Air Purifiers

| *Botanical Name* | *Common Names* | *What they filter* |
|---|---|---|
| *Sansevieria trifaciata* | Snake plant/ 'mother-in-Law's tongue' | An ideal bedroom plant as this not only removes volatile organic chemicals like benzene, formaldehyde, xylene and trichloroethylene but also replenishes oxygen at night times too. |
| *Spathiphyllum wallisii* | Peace lilies | Remove benzene, carbon mono oxide, trichloroethylene and formaldehyde. Helps to keep shower tiles and curtains free from mildew and absorb harmful vapours from alcohol and acetone. |

| *Botanical Name* | *Common Names* | *What they filter* |
| --- | --- | --- |
| *Syngonium podophyllum* | Decorative plant available in variety of colours | Acts as air purifiers, remove volatile organic chemicals, improve humidity and kill air borne microbes. |
| *Chamaedorea seifrizii* | Bamboo palm/ cane palm/reed palm | Filter benzene,formaldehyde and trichloroethylene. |
| *Phoenix roebelenii* | Pygmy date palm | Capable of removing formaldehyde. |
| *Nephrolepis exaltata* | Boston fern/ sword fern/ fishbone fern | Filter formaldehyde, benzene and xylene emitted from many home furniture and building materials. |
| *Scindapsus aures* | Golden pothos/devil's ivy | Tackle formaldehyde, carbon mono oxide and benzene from indoor air. |
| *Gerbera jamesonii* | Gerber daisies | Effective at removing trichloroethylene, which we may bring home with dry cleaning. Also good for filtering out benzene that comes with inks. This powerhouse plant helps to remove all indoor toxins. |
| *Chrysanthemum morifolium* | Pot mum | Help to filter out benzene commonly found in glue, paint, plastics and detergent. It is air purifying champion, removing ammonia, formaldehyde, xylene from indoor air. |
| *Aloe vera* | Ghrit kumari/ ghee kuwar | Keeps the home free from benzene which is commonly found in paint and certain chemical cleaner. |
| *Hedera helix* | English Ivy | Reduce the amount of fecal matter, absorb formaldehyde from cleaning products and trace amounts of benzene found in office equipments. |
| *Dracaena marginata* | Red-edged dracaena | Best for removing xylene, trichloroethylene and formaldehyde introduced through lacquers, varnishes and gasoline. |
| *Chlorophytum comosum* | Spider plant | Works hard to remove the air of harmful pollutants like formaldehyde and benzene. It is also a carbon monoxide eliminator. |

# Chapter-12

# Foods Polluted by Chemicals Cause Havoc

We don't just get sick but are being made sick. Why we and why now? Do have answers. YES!

Advancing Technology carries an expensive price tag. One of this price tag is toxic chemicals in food and water. Humans are exposed to chemicals through food we eat, the air we breathe and water we drink and bathe in. Many chronic diseases in our body spring from these toxins. There are potentially harmful or toxic chemicals present in food we eat whether occurring naturally as contaminants or as deliberate additives. They may enter food supply during growing, processing and packaging. These include pesticides applied to crops, hormones, and other substances included in animal feed etc. Some toxic chemicals in food are added to combat impurities and bacteria. Others are added to enhance flavor and give 'eye' appeal to food products as well as to increase the shelf life of foods. These additives, preservatives and colours used cause slow or even rapid poisoning once ingested. Some toxins are created during food processing.

Ordinary fruits and vegetables are expected to be edible once plucked from farm. Not so again. Industrial pollution and attendant chemical settlement that are invisible to the eye constitute great health hazards that threaten human lives.

## Chemicals Present in Foods

Toxic chemicals may be present in the food in number of ways-

## 1. Additives

They are added to food in order to provide some useful purpose such as flavor and preservation which allow consumer to select a varied diet from preserved foods all the year round.

Today's hyperactive children are the result of all artificial colours they ingest with fruit juices, soft drinks, ice creams and fruit jams. There is enough evidence to suggest that food additives and artificial colours cause harm to health and children are worse affected. A study of students in New York showed that students who ate lunches that did not include artificial flavor, preservatives and dyes did 14% better on IQ Test than students who ate lunches with these additives. Food additives are linked to tantrums in 25% of toddlers. Adults are also adversely affected. In several countries, use of food colours like sunset yellow is either banned or regulated but not so in India. There is suspected link between food additives and changes in children's mood and behavior. Fruit juices contain additives, preservatives, artificial sweetener and colours. A study published in The Lancet in November 2008, looked at the effects of fruit juice additives on children behavior finding that " Artificial colour or Sodium benzoate or both" in diet result in increased hyperactivity in 3 years and 8-9 years old children in general population.

Colours are added to food for cosmetic reasons- to make preserved food look colourful and more appealing to consumers or to restore natural colour that has been lost during processing and storage, for example- red chili powder with red Sudan dye, red colour in rose syrup, green in khus syrup etc. In summer, streets are lined with water melon sellers who inject red dye with saccharine to give fruit a bright red shade and sweetness making them toxic. Besides, okra, peas, capsicum and brinjals are some vegetables given a colour bath or shiny coat. Other foods like pulses and spices are coated with a paint to give them attractive shine.

The different colours used are-

*Carmosine (Azurubine)*- Synthetic red dye with International Numbering System (INS) E-122. It appears to cause allergy, rash and skin swelling, provoke asthma.

*Sunset Yellow*- Orange yellow dye with INS E- 110. It is a carcinogen, causes gastric upset, vomiting, nettle rash (urticaria), hyperactivity in young children.

*Tatrazine*- Lemon yellow dye with INS E-102. It causes anxiety, migraine, itching, hot flashes, sleep disturbances etc. Studies have linked it to childhood obsessive compulsive disorder and hyperactivity.

Once approved by Food and Drug Administration (FDA), food additives are considered fit for human consumption- but they may not be entirely safe. Some food and colour additives have induced allergic reactions while others have been linked to cancer, asthma and birth defects. Hundreds of years ago before the advent of food additives, chemically polluted drugs, refined sugar products and preserved foods, these disorders did not exist.

## 2. Ripening Agents

Besides chemical cuisine, the natural fruits nowadays have become un-natural due to artificial ripening agents. Markets of each and every city is flooded with Calcium carbide ripened mangoes, bananas, alcohol injected watermelons, saccharine filled muskmelons. Eating artificially ripened fruits may adversely affect one's health even leading to cancer. They are extremely hazardous because it contains traces of arsenic and phosphorus. When calcium carbide comes in contact with moisture, it produces acetylene gas which is analogous to ethylene present in fruits for natural ripening. Nowadays, apples are coated with thin film of wax to provide them a shiny appeal as well as to increase their shelf life.

This acetylene gas causes headache, dizziness, mood disturbances, memory loss, and cerebral oedema. Five radicals from carbide play a major role in ageing process, cancer and heart diseases, stroke, arthritis and various allergies. When it is mixed with oxygen, it acts as a sedative. It is used in gas welding of steel goods.

## 3. Hormones

Nowadays farmers use oxytocin liquor to promote growth of fruits and vegetables like cucumber, gourds, melons, and apples etc. making fruits hazardous and tasteless. Oxytocin is a mammalian hormone that acts primarily as a neurotransmitter in the brain. Commercial milkmen inject it to their cows and buffaloes to increase the quantity of milk. The milk may contain the traces of it. It affects the reproductive capability of females. Nausea, cramps, vomiting, stomach pain may occur after the excess use of oxytocin in fruits and vegetables.

## 4. Junk Food Additive

The junk and all Chinese foods are flavored with additive like Mono sodium glutamate (MSG) with E number 621 commonly known as "Azinomoto". It masks off flavor and makes the blandest and cheapest food taste wonderful. MSG is a neurotransmitter that excites our neurons (not just in our tongue). Free glutamic acid from food sources can get into brain injuring and sometimes killing neurons. Excessive consumption may lead to loss of memory, hyperactivity, brain damages, numbness, stroke, difficult breathing, rapid heartbeat, headache, nausea etc. The terms used on labels "natural flavour" and "hydrolyzed yeast extract' are often forms of MSG.

## 5. Pesticides

Persistent organic pollutants (POPs) residues are present in virtually all categories of food including backed goods, fruits, vegetables, meat, poultry and dairy products. Most commonly found POPs being the pesticide DDT and dieldrin. Large DDT residues in human tissues and human milk were reported from the consumption of food containing traces of DDT. It also reaches human foetus by crossing the placenta. Potential mechanism of DDT action on humans is genotoxicity and endocrine disruption. A Japanese study concluded that DDT exposure may affect thyroid hormone levels causing hypothyroidism. Studies on human population indicate that DDT causes cancers of liver, pancreas, breast and testicular etc.

## 6. Chemicals in Food Chain

Substance that resists chemically or breakdown accumulates as it is ingested by one species after another steadily building up as the food chain progresses from small weak species to large and dominant. The highest level of pollutants, therefore, is ingested by large animals at the top of the food chain. These toxic chemicals enter the human body and cause adverse health effects. Whether a contaminant is harmful or not depends on how long it lingers in the body or the environment. If the body rapidly excretes a substance or metabolizes it into harmless compound, brief exposure may not harm. But when a contaminant interacts with body system, it may be dangerous. It becomes dangerous when body transforms them into reactive compound that can damage its DNA. Some toxins are stored in fatty tissues and remain out of harm way until the fat is mobilized from energy. These toxics may cause a number of health problems like – direct damage to lungs, liver, kidney, bones, blood, brain, nerves and reproductive system, cancer, asthma, high BP, Parkinson's disease,infertility,genital malformation, deficient in attention memory learning etc.

Early in 2008 in China, an abnormal increase in infant cases of kidney stones makes people to look at China milk poisoning incidents. It was due to powder milk mixed with 'Melamine', an industrial chemical used for production of melawares, home decoration and in industrial production. Why is this melamine powder added to milk powder? Melamine has same protein (containing nitrogen) as in milk. It does not have any smell. Adding melamine into milk reduces the actual milk content required giving more profit to traders. Kidneys retain the melamine forming stones; block the tubes causing problems in urination. Ultimately, it leads to loss of kidney function and will require kidney dialysis or lead to death. Baby's kidneys are very small and they drink a lot of milk powder and thus suffered from stones.

## 7. Natural Components

Some natural components of plants may themselves cause toxicity e.g. glycoalkaloids in potato while some may be harmful if not cooked properly, for example- lectins in pulses. Large amount of acrylamide in food rich in starch cooked at high temperature mainly potato and cereal based products such as chips, crisps etc., is known to cause cancer in animals and peoples.

According to Dr. Mercola in his recent article - ' Is something wrong with our Modern Diet'. He opined that most common grocery stores, restaurants and vending machines are continuing to mainly sell heavily processed foods and pre-packaged products. These products often contain high levels of sodium, sugars, unhealthy fats, chemical additives that cause health problems. There are number of technologies that food manufacturers can use behind the scenes and mainly on labels that will give products the fresh-like quality. Everything related to naturalness and freshness is being manipulated constantly by the manufacturers. To ensure sound nutrition while guarding against exposure to high levels of pollution and pesticides, the wisest course is to eat a wide variety of foods and avoid consuming too much of a food items that may carry heavy contaminant load. For contaminants we can't avoid, our bodies are remarkably well equipped with preventive mechanism to detoxify them. To stay healthy and energized, we have to shift our diet from processed foods to more natural and home cooked versions.

## What Can We Do?

- ☆ Elimination of sugar, colouring, flavorings and preservatives from the diet.
- ☆ Follow a healthy eating plan.
- ☆ Good quality of mineral supplement and omega 3- oil for every child.

- ☆ Avoidance of fats, poultry.
- ☆ Avoidance of junk foods.
- ☆ Foods from dairy products/ milk powder should be avoided.

## We Should Consume Plenty of-

- ☆ Broccoli and other cruciferous vegetables for cancer fighting indoles.
- ☆ Vegetables and seasonal fruits for detoxifying bioflavonoides.
- ☆ Fresh fruits and vegetables for antioxidants, nutrients.
- ☆ Fish and vegetable oil for omega 3-fatty acids.
- ☆ High calcium foods abundant in dairy products as well as dark green leafy vegetables to reduce intestinal tumor growth.
- ☆ Onion and garlic to reduce exposure to carcinogens.

Luckily, the plants also have cancer fighting substances that cancel out the potentially harmful effects of others. The food that we choose matters. By making just a few small changes in our eating and buying habits, we may greatly reduce the impact of our food choices on the environment and ourselves. The motivation for consumption of "ORGANIC FOOD" is needed because they are natural (without any sprayed chemical), nutritious and tasteful. They have been now accepted due to their perceived health benefits over conventional foods.

*"Let healthy foods be your medicine rather than medicine your food."*

# Chapter-13

# Food Security: A Global Challenge

Everybody needs food. Food fuels life. Life is precious. Our planet is now facing issues like hunger, health, growing population, need to form sustainability etc. Everyday, there are 2,00,000 more mouths to feed. By 2050, there will be about 2 billion more people who will need food when climate change would have also its impact on uncertainty of production and even reducing the essential output. The demand for food will be 60% more than it is today. In the next 50 years, world has to produce more food than in the past years. Thus, the complexity of delivering sufficient food to nation's population as well as whole world's population shows why food security matters for all countries, whether developing or developed. Food security directly relate to nutrition and health. Typically food security is thought of as being related to availability of food and one access to it. A house hold is considered food secure when its occupants do not live in starvation or hunger. Food security is the fundamental human right. It includes both physical as well as economic access to food. It is a multidimensional concept.

The World Food Summit, 1996, defined food security as- 'a situation that exists when all people at all times, have physical, social and economic access to sufficient, safe, and nutritious food that meets their dietary needs and food preferences for an active and healthy life'. Food security is not defined as whether food is available or not, but whether the monetary and non- monetary resources are at the disposal of the population so as to allow everyone to access to required quantity and quality of food. Food security will not only depend on climate and socio-economic factors for food production, but also on economic growth, changes in trade flows stock and food policy. There are many factors in today's global environment that exacerbate food security. It is true that we are living in an age where we are growing and producing more food than ever before. During the last few decades, the total food production in India has increased at a much faster rate than the population. Green revolution started in late 1967-68 was a noteworthy watershed resulting in phenomenal increase in production of crops especially in food grains that has changed the food security condition in India. It is important to understand the role of science and technology in ushering the Green Revolution, which ensured food security in India. Similarly today, innovations in biotechnology may be a game changer in battle for nutrition security. We have enough food to feed the world's population but it is not distributed properly nor is all food culturally appropriate across the globe. Food security is the outcome of food system. A food system is a dynamic interaction between and within bio-geographical and people's environments that affect activities and outcomes all along the food chains. Multiple socio-economic and bio- physical factors affect food systems and hence food security. In India, Food security is embodied under Directive Principle of State Policy (DPSP) as exemplified in constitution of India, 1950.

## What Causes It ?

The various reasons are-

**1. Population Growth:** The population varies considerably across the countries. It in developing countries are increasing day by day. Population pressures and land degradation undermines current  and future attempts to boost food yields.

**2. Changing Tastes:**  Not only the population is growing but its food diets are changing too. They have started eating foods that are richer in processed food, meat and dairy products. But to produce more meat and dairy milk etc. means growing more grain.

**3. Climate Change:** Rising temperature will turn arid zone into more deserts. Current assessment of climate change impact indicate that some regions may be benefited by increased agriculture production while other may have to face reduction in yield depending on their location. At current rates, amount of food we are growing today will feed only half of the population by 2050.

**4.Water Scarcity:** Agriculture is important for food security. But 28% of agriculture lies in water stressed regions. For both rainfed and irrigated agriculture, the variation in rainfall is the deriving element. The short time variability of rainfall becomes the major risk factor. Due to declining ground water level and rising sea level, the aridity of the soil or increasing level of saline is on increase reducing the suitability of land for cultivation of crops.

**5. Troubled Farmers:** Fewer and fewer people are choosing farming as an occupation. Meanwhile food prices are hiking. Soil fertility is being degraded due to various reasons.

The variability in four components of food security makes the food system vulnerable. These components are-

## 1. Food Production and Availability

It refers to 'supply side' of food security. Increased availability of food is an essential condition for achieving food security in India. Food availability relates to supply of food through production, distribution, carryover stocks and imports. The climate change directly affects food production through changes in afro-ecological conditions and indirectly, through growth and distribution of income and thus demand for agricultural produce. Availability is declined by the physical quantity of food that are produced, stored, processed, distributed and exchanged. Food security directly depends upon the net availability of food grains including the price of it.

## 2. Stability of Food Supplies

Greater fluctuations of crop yields due to severity of extreme weather conditions and local food supplies can affect adversely the stability of food supplies and food security.

## 3. Access to Food

Access to food refers to the ability of individuals, communities and countries to provide food in sufficient quantity and quality. It is a measure of ability to secure the resources including legal, political, economic and social that an individual and/or society requires to obtain access to food. Falling real prices for food and rising real incomes over last three decades have led to substantial improvements in access to food in many developing countries. Possibility of increase in food price and decrease in income growth rate due to climatic change may reverse the trend.

## 4. Food Utilization

Food utilization refers to the use of food. It includes and covers the nutritional value of food taken as diet.

Achieving food security under changing climate requires a substantial increase in food production on one hand and improved access to nutritional food and capacities to cope with the risk posed by climate change on the other.

# Environment and Food Security

Natural resources are fundamental assets to the production of food, rural development, sustainable growth and population well being. The pressure on natural resources in various regions of the world is increasing as their most efficient use, conservation and containment of negative impacts are tied to the process of economic development. Climate change shows its significant impact on agriculture and its ability to provide food production. These impacts may be both 'direct'- impact on biophysical processes and on agro-economic conditions at the agricultural system ; as well as 'indirect'- impacting growth, wealth distribution and also on demand for agriculture products. Therefore, climate change will have negative effects on global food security, thereby increasing dependence of developing countries on import and increasing the precarious conditions of a number of people.

It is estimated that decline and conservation of crop plants use could cause reduction in cultivated land area of 8-20% by 2050. The cumulative impact of climate change, land degradation, crop land losses, water scarcity and species infestations may cause projected yields to be 5-25% short of demand by 2050.

## Society and Food Security

Social aspects of food security are based on three interrelated areas- Human Health, Demography and Socio-political issues.

The relationship between human health and food security is an issue in developing countries. Climate change may cause a decrease in food security in some developing countries. Malnutrition is a global health problem. The malnutrition and undernutrition affect the immune system of individuals, their exposure to diseases and severity of the disease. The FAO report on food insecurity (1999) identified population groups, countries and regions which are especially vulnerable. For example, nearly half the population of countries in central, southern and east Africa is undernourished. Undernutrition is a fundamental cause of stunted physical and mental growth in children, of low productivity in adults and of susceptibility to infectious disease in everyone. In addition to this, there are so many social- economic conditions that emphasise the linkage between disease and malnutrition, such as unfitness to work, lack of nutritional knowledge etc.

The complex chain of events involving food and/ or diseases both is influenced by many other social- economic variables that make the scenario more complex and multifaceted. The various social- economic variables may be – education, living conditions, food prices, food availability, health and hygiene, social political stability, Population explosion and urbanization in developing countries etc. Displacement due to longer- term cumulative environmental deterioration ( for example, land degradation and water scarcity) is also associated with many health impacts.

It seems that in future, there may be a risk of worsening in availability and security of agriculture and food products leading to increase in level of social conflict,  particularly in developing  regions where food and water are  incredible multiplier factor.

## Challenges of Food Security

India faces challenges to meets its 'Food Security' is as follows-

## Crop Diversification

The prices of food grains like rice and wheat are not promising and encouraging to farmers. By concentrating on other crop, farmers may be encouraged to earn more profit.

## Climate Change

There is climate change due to rising temperature and extreme events on food production system which affects agricultural growth adversely. Climate change is also expected to impact agricultural land use and production due to less availability of water for irrigation and other factors.

## Mismatch between Water Availability and Demand

In India, there is variation in pattern of precipitation. Water demand is for various purposes is increasing due to population explosion, urbanization and industrialization. The agriculture sector, at present,  is using about 83% of water resources but availability of water may decline to 68%  in 2050 due to  demand from other sectors. Therefore, mismatch between water demand and supply  may negatively affect the food grain production.

## Agricultural Marketing

Agricultural marketing and agricultural produce should be improved by involving  private sector and Foreign Direct Investment in India.

## Land Fragmentation

Low level of agricultural productivity may be result of land fragmentation. Due to  population explosion, the agricultural land is converted into non-  agricultural leading to low productivity, thereby creating  food insecurity. Therefore, today's need is to make changes in land use and cropping system.

## Quality Seeds

It is essential to make available quality seeds with good genetic potential to the farmers at reasonable and affordable prices.  So that these farmers become enable to harvest more yield.

## Globalization

It has brought changes in technology, development, transport, communication posing challenges among farmers as well as in society. Globalization has intensified interdependence and competition between economies in the world market. Therefore, farmers have to face worse conditions.

Performance, challenges and policies in food security in India have been examined in terms of availability, accessibility and utilization. All these three terms are interrelated. Despite India's journey  to ensure food security, it is still in a pathetic state. India's food security status remains to rank as 'alarming' according to International Food Policy Research Institutes's Global Hunger Index, 2014. It ranks 55 of the 120 countries in the world. Notwithstanding its food grain surpluses, India faces a challenge of nutritional security. As per 2018 Global Hunger Index (GHI) published on October10, 2018,  the level of hunger and undernutrition worldwide fell to 20.9 down from 29.2 in the year 2000. As per the index, India was ranked 103rd out of 119 qualifying countries. FAO's recent publication, The State of Food Security and Nutrition in the World, 2018 estimates that about 15% of Indian population is undernourished. Due to inflation, the cost of food items is increasing day by day leading to hardships. In addition, land degradation and fragmentation, crop diversification, climate change, unavailability of irrigation facilities have added up to woes of producers.

Government and its policies play a major role in food security. These policies contain Public Distribution System (PDS), Employment Generation Scheme, Social Protection Programme, National Security Bill, Antyodaya Yojana etc. Achieving 'Food Security' is a far cry for India, even though India is self sufficient in growing food grains.

# Chapter-14

# Natural Resources in 21$^{st}$ Century

Ever since the birth of human beings on planet earth, man has been completely dependent on natural resources around him for his survival and basic needs. The story of evolution of civilization is the story of acquisition of knowledge regarding these natural resources available on earth and their utilization for human welfare. Natural resources are resources that occur within environment that exist undisturbed by mankind, in a natural form. On this planet, it includes sunlight, atmosphere, water, land ( includes all minerals) along with all vegetation, crops and animal life that naturally exists. Many of the natural resources are essential for our survival while others are used for satisfying our needs. These natural resources are materials that can be found within the environment. Every man made product is composed of natural resources. These may exist as a separate entity (such as fresh water, air and living organism like fish) or it may be in an alternate form that must be processed to obtain the resource such as metal ores, rare earth metals, petroleum and most form of energy. The natural resources are generally of two types-

## Renewable Resources

Renewable resources can be replenished naturally. Some of these resources like sunlight, air, water etc. are continuously available and not affected by human consumption.

## Non-Renewable Resources

Minerals are the most common resource of this category. Such type of resources are non-renewable, when the rate of consumption exceeds the rate of recovery. The best example is fossil fuel. Some of these resources naturally deplete in amount without human interference, for example- radioactive elements such as uranium which naturally decay into heavy metals.

Abundantly endowed by nature, Indian land is blessed with diversity in climate and heritage that has nourished us through the ages. India has rich resources of land, forests and water. This trinity of land (Jameen), forests (Jangal) and water (Jal) are interdependent and interrelated and determine the kind of livelihood support system for people particularly in rural areas. We humans have built our tradition, culture, consumption, habitat pattern and also livelihood around these natural resources. The interrelationship of natural resources and people are a part of a complex web in which socio-cultural values, health, knowledge, life style and environment are interwined.

## Land

Land is an essential natural resource, both for survival and prosperity of man and for maintenance of all terrestrial ecosystems. Land may be referred as dry land/ solid surface of earth that is not permanently covered by water. It occupies nearly 20% of earth surface. Most of the human activities throughout history have occurred in land areas that support agriculture, habitat and livelihood etc. It provides food, fiber and medicine to people. Human beings, therefore, use land resource for production, residence and recreation. Land also acts as a dustbin for most of wastes created by modern society.

In the current century, people have become progressively more expert in exploiting land resource for fulfilling their own need and greed. The limits on these resources are finite but human demands are not. Increased demand and consumerism tendency or pressure on land resource results in declining crop production, degradation of land quality and quantity and competition for land. Major impact of land use change by people includes sprawl, soil erosion, soil degradation, salinization, desertification, urbanization, and environmental pollution etc. Currently, land resources are under stress.

## Forest

Forest includes all uncultivated and uninhabited land. Today, forest is any land managed for the diverse purpose of forestry whether covered with trees, shrubs, climbers etc. In Indian subcontinent the term forest does not merely mean an area covered with trees but it carries the impression of an entity that is sum total of ecological, edaphic and biological parameters. Forest ecosystem is a biotic community usually with a closed canopy and the physical environment in which trees, dominant life forms and micro flora and fauna find shelter. Forests are a renewable source and contribute to economic development. They have a major role in enhancing quality of environment. Forests are our precious assets. They are home to much of the global biodiversity, store huge amount of carbon, and have significant role in reducing the vulnerability of rural people who depend on forests for their livelihood- food, fuel, fiber, medicine, income etc. In country like India, forests were omnipresent. These public forests have to meet multifarious demands of we people. But today, the scenario has totally changed. Our natural forests cover is being rapidly degraded and deteriorated. At present, forests form just 20% of landmass with their quality deteriorating day by day. Now, the forests are at the mercy of people who are exploiting and destroying the green cover at an alarming rate.

The forests play a very significant role in human life. They act as earth's lungs supplying the oxygen required for survival of living beings. Forests regulate rainfall and earth's climate. They absorb most of the pollutants of the atmosphere acting as carbon sinks as well as intercept sound, thus reducing noise decibels. Therefore, deforestation is responsible for 1/5th of greenhouse gas emissions. Population growth and its pressure are closely linked with deforestation. More population means more food for more mouths. And there is demand for more land for agriculture resulting in the conversion of forest land into agriculture land. Blind infrastructure development of roads, dams, buildings, power plants, mining, railway lines and residential societies etc. directly affect forests negatively. Now in view of present scenario, conservation and sustainable development of forests is the only solution.

## Water

Water occupies a pivotal position among the geo resources of the planet earth. In fact earth is perhaps the only planet endowed with water in abundance. It is a bond between human beings and nature. In the social and religious traditions and culture of India since the Vedic Times, water has enjoying the most respectable and unique status. The life owes its origin and sustenance on the presence and availability of water. Water came to be regarded as precious and its conservation and preservation was sanctified by religion. Our history reflects the wisdom of our forefathers who made harvesting of water and its management an integral part of native culture and community life. These practices were sacred duty of the people, therefore, no water problem in those days. Owing to increasing industrialization on one hand and exploding population on the other, demands of water supply have been increasing at an alarming rate. Water quality issues are a major challenge that humanity is facing in 21st century. Geological factors, increasing population, rapid urbanization, agricultural developments and poor waste water regulation have affected the quantity and the quality of water. Fresh water resources all over the world are threatened not only due to over exploitation and poor management but also by ecological degradation. Deforestation and mining have destroyed the ability of water catchment to retain water. Since the middle of 20th century and concurrent with onset of accelerated industrial growth, various types of water pollution problems have occurred in rapid succession. Major causes of water pollution are organic sewage, detergents, agricultural chemicals, industrial effluents, silts, oil deposition, and radioactive substances in the water bodies. India's water crisis is predominately a man made problem. If all this continues, then very soon i.e. by 2030, nearly half of world's population will be facing water shortage problem and will not have to access to sufficient drinking water. It is predicted that in 21st century, there will be 'Water Wars'. The day is not far when we have to pay heavy price to buy the safe drinking water (still we are buying the drinking water bottles). Now our focus should be on changing water consumption patterns to avoid over exploitation and wastage so as to conserve country's precious resource.

But during past few decades, over exploitation of environment and natural resources is becoming increasingly widespread in modern world. Human greed has taking its toll in form of degraded environment and fragile ecosystems. The over exploitation of natural resources is a key factor in economic growth and associated development, but it has negative impact on environment. It results in the destruction and degradation of forests, the depletion and pollution of water resources, decimation of fisheries, degradation of land and extraction of mineral resources. Population explosion is acting as a catalyst for resource depletion. Urbanization, globalization, technology advancement, industrialization, new agricultural practices, increasing trend of tourism, increasing transportation and natural calamities have become another drivers  for environmental degradation and natural resources depletion. Now, these three above said major resources are fading fast. The data available point out that earth is now reaching its limits in the use of its natural resources. Today due to current rate of consumption of natural resources, the demand exceeds 41% spare capacity of the earth. By 2050, when global population will be more than 10 billion, our earth cannot exist to meet such demands of natural resources.

An estimated 60% of cultivated land suffers from soil erosion, water logging, salinity and 5-10 billion tons of top soil is lost annually from soil erosion. Ground water tables are falling fast and forest cover is under threat. The rapid development of new transportation links has taken place without a parallel increase in government capacity to protect natural resources such as forests. Improved roads significantly raise opportunities for increased illegal and legal logging and mining. Many hydropower dams are built to provide power for mining and industry. The dams destroy forests and watersheds and both the dams and mines pollute the rivers. Shrinking supplies of water and land subsidence in many cities are the result of allowing the destruction of forested watersheds. Available data on reserves of mineral resources is now reaching its limit. According to Fernando Alcoforado, its predictable extinction and consumption are-

- ☆ *Platinum-* Useful in surgical materials , extinct by 2049;
- ☆ *Silver-* used in mirrors and cutlery, extinct by 2020;
- ☆ *Copper-* Used in wire and cables and A.C.ducts, extinct by 2027;
- ☆ *Lithium-* in cell phone batteries, laptops etc.,  extinct  by 2053;
- ☆ *Phosphorus-* in agricultural fertilizers, extinct by 2149;
- ☆ *Uranium-* in electric power generation, extinct by 2026;
- ☆ *Titanium-* in camera lenses, extinct by 2027;
- ☆ *Nickel-* in metal alloy coating in electronic, extinct by 2064;
- ☆ *Tin –* in coating metal alloys, for example- soft drink cans by 2024;
- ☆ *Gold-* in jewelry and computer microchips, by 2043 etc.

The current pattern of natural resource exploitation is environmentally destructive, socially inequitable and contributes to human insecurity, political instability and social conflict. Globalization and supply chains of developed countries and rapidly growing economies, such as China and India, are drivers of unsustainable development and environmental, socioeconomic and related transboundary impacts such as inequitable use of common resources. Unchecked population increase with urbanization in many parts of the globe has resulted in excessive consumption patterns, rapid degradation and depletion of natural resources like water and forests.

Depletion of natural resources is considered to be a sustainable development. It is balancing the needs of people and species on earth now and in the future. Depletion is also associated with social inequality. According to Nelson, depletion of natural resources is caused by 'direct drivers of change' such as mining, petroleum extraction, fishing and forestry, as well as 'indirect drivers of change' such as demography, economy, society, politics and technology. The current practices of agriculture are another factor causing depletion of natural resources, for example- the depletion of nutrients in soil due to excessive use of nitrogen and desertification. The depletion of natural resources is a continuing concern for society. The sustainable development and natural resources have permanent interdependence of society existence, of development of the economic, technologic social human medium and of ensuring the environment protection as shown in Fig.4.1 given below.

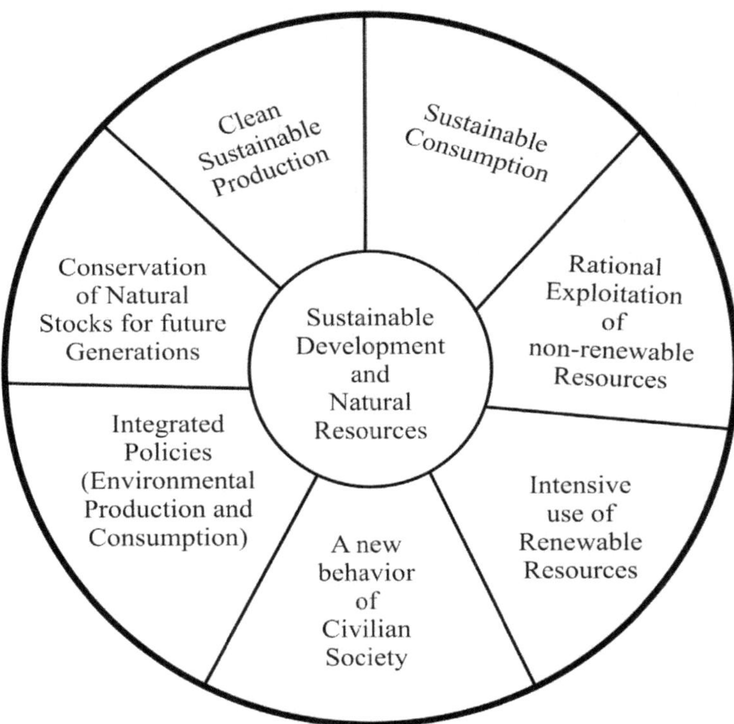


**Fig. 4.1: Sustainable Development and Natural Resources Interdependencies**

There is dire need to take measures at all societal levels to protect nature from further depletion due to human activity. There should be sustainable use of natural resources and protection of resources should be incorporated into national and international systems of law. Ecological footprint is the method used to measure the amounts of land and water in terms of global hectare that would be needed to support domestic consumption. The ecological footprint is a calculation of what each person, every country and ultimately the world's population consume the natural resources. Community based resource management is looking as the perfect solution. Recognizing the interrelationship and interdependence of water, forests and minerals is essential in managing the resources and reducing environmental deterioration and impacts of natural disasters. After facing the consequences of exploitation of natural resources in form of loss of livelihood, disasters, or poor health conditions, people are inculcating awareness that deforestation may led to severe flood while excessive mining may result in water pollution.

The government policy makers, Non Government Organizations'(NGO), local communities, environmentalists and scientists and the media should communicate with one another and have to develop a common understanding of interrelationship between natural resources and human security for sustainable use of natural resources, limiting further environmental degradation and natural disasters. Long term environmental sustainability may be pursued through changes in the education system that can foster more positive behaviour and attitudes towards surrounding environment and use of natural resources. Besides, tree plantations and other greenery is very essential in order to preserve natural resources for better livelihood of individuals, animals, industries, agricultural crops etc. Economic impact of green development is always positive. India is a developing country with increase in industries and development of technology, innovation and other advancements, it is necessary to implement the measures to check all kinds of pollution, plant more trees and encourage greenery and follow waste management in order to preserve the natural resources.

# Chapter-15

# Anthropogenic Impact on Planet Earth

India being one of the 12 mega diversity countries in the world is rich in floral and faunal wealth. India has the potentiality of harnessing its bioresources for health, industry and livelihood. Natural environment is of crucial importance for social and economic life of us. Nature does not need us to rule over it, but runs itself very well and better without humans. We are the parasites on food chain of life, consuming more and more and giving too little back to restore and maintain the life system that support us. We, humans utilize the environment as-

☆   A resource for food

☆   An energy source

☆   A source for recreation

☆   Natural resources for industrial product

☆   A major source of medicine

Therefore, biodiversity of nature not only fulfills the current needs and desires of man but also foster the role of nature as a source of solution for future needs and challenges. Primitive man used natural resources to satisfy his basic needs of air, water, food and shelter. These unprocessed natural resources were easily available in the biosphere. The residues left after use of these resources were easily and directly assimilated by the environment.

However, only humans have the potential to accumulate resources from beyond their immediate surroundings and process them into varied versatile forms. This has made man to flourish beyond natural constrains. With the dawn of industrial revolution, man was better able than ever to satisfy his needs of air, water, food and shelter. So, he diverted his action to other needs beyond those associated with survival. These acquired needs are usually fulfilled by processed, manufactured or refined products. As a result, anthropogenic (man-induced) pollutants have overloaded the system disturbing the natural equilibrium. Technological development has resulted in more and more exploitation and depletion of natural resources, mainly coal and petroleum and various minerals. Mining activities, road construction, urbanization and industrialization have interfered with ecological balance.

Today, however, human pressure on natural environment is more than before in terms of magnitude and efficiency in disrupting nature and natural landscapes. Due to pressure on land, water and bio-resources, there is continuous impact of various types of human activities on our environment in one form or the other. The modern polluted environment is the mirror of human's undesirable activities.

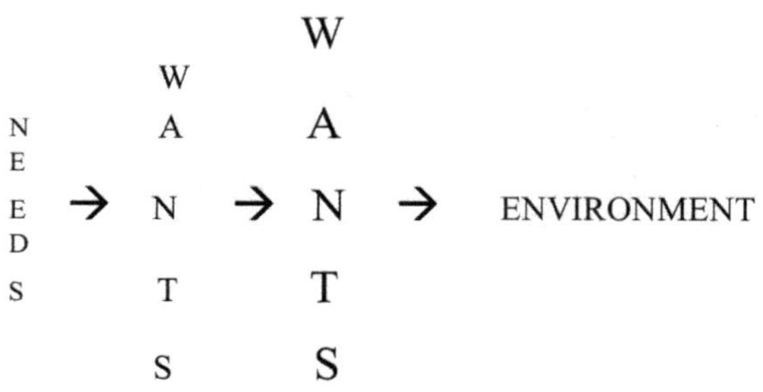


**Fig. 5.1: Social needs and wants**

## How do Humans Affect the Environment?

The ways in which human beings have changed and are changing the face of earth, the degradation is due to-

### Agriculture

Human civilization is dependent on agriculture. It traditionally depends on sound environmental conditions. As a result of needs for more food production since 1940s, policies have implemented for increased production through modern techniques. Modern agriculture means manipulating the environment to favour the plant species (crop) that we can eat. With this agriculture, we are creating a very simple ecosystem. It has only three levels- producers (crops), Primary consumers (livestock and human) and secondary consumers (humans). This is good for us but what type of 'ecosystem' have we created?

Agriculture ecosystem has many problems. First, the highly specialized monoculture fields (with only one crop), it packs many similar plants into a small area, creating favourable condition for disease and insect pests. In natural ecosystem, plants of one species are often scattered. Because insects have got specialization on feeding on particular plant species, they have to face hard time finding the scattered plant species. Ultimately without food, the insect populations are kept under check. Disease is more easily spread in plants with close proximity. To keep monoculture going, it requires a lot of pesticides. Other problem with this agriculture is that people have to depend on few crops for food. If suppose one crop fails worldwide in same year, we should have to face hardship to feed everyone. Natural ecosystems have alternative crops for food. Nutrient recycling is another problem of modern agriculture. In natural process, when plant dies, it falls on the ground and its inorganic nutrients are returned to the soil after rotting. In agro-agriculture, the crop is harvested, trucked away and finally it is flushed to be run off into the river creating water pollution. Therefore, nutrients are not returned to the soil, besides they are replaced by chemical fertilizers. These chemical fertilizers tend to run off the fields and further pollute the water. Besides, monoculture is causing loss in species abundance and diversity.

## Energy

All energy types have potential impacts on natural environment at all stages of use from extraction through processing to end use. For example- Thermal power plants. These plants need fossil fuels including coal, diesel and natural gas. Mostly are coal based and hence, is the dominant contributor to problems with energy generation and environment. To generate one unit of energy, an enormous amount of coal has to be burnt under controlled conditions. Most of our coal deposits are located in Eastern India especially in Jharkhand and Bihar. Several steps and points of conflict from the resource to the end point in power generation and environment can be listed, such as acid mine drainage, ash generation in power generation set up, exploitation of many mineral resources like marble, metals etc. leading to various types of environmental pollution etc.

## Fisheries

This sector is in a state of crisis with over capacity of fleet, overexploitation of stocks, debt and marketing problems. The fishing activity has an adverse impact on cetaceans and large number of dolphins and even the globally endangered monk seal are being killed.

## Industries

The impact of industry varies at different stages in processing of a product, depending upon the raw material used through to the final end use of the product for waste residue, re-use or recycling.

## Transport and Infrastructure

Nowadays, transport is the major contributor to pollution and green house effect globally. The key impacts of transportation are- habitat fragmentation, disruption of migration and traffic moralities of wildlife. Since 1970, transport has become a major consumer of non-renewable resource, 80% of oil consumption coming from road transport.

## Tourism

Tourism in new millennium is India's largest industry providing 15% of foreign exchange earnings. Wild life tourism is the fastest growing new concept. Instead, it contributes to environmental degradation and loss of natural resources. Construction of hotels, refuse leftovers, sewage water disposal in water, air pollution due to transport and tourist vehicles, noise pollution by recreational vehicles, scuba diving and insensitive tourists damage the rare flora of medicinal importance.

## Technology

Though technology is making lives of man easier and comfortable. It poses a great threat to environment. Radiation hazards are increasing day by day due to use of mobile and Wi Fi around us. Hence, we can notice that many small birds and insects like honey bees are not found around these days.

## Non- vegetarian Diet

Human digestive tract is a long one like that of the herbivorous animal. But most of the human population relies on the non-vegetative diet. This reliance on non-vegetation diets expensive in terms of environment. Because to grow a hen of 1 kilo weight, we need many kilos of wheat. Instead, a kilo of wheat is sufficient for a diet of two individuals. So, we grow animals for food at the cost of many kilos of herbal diet. It requires growing cereals in many acres of use of fertilizers, pesticides etc. which in turn pollute the environment. Similarly, we kill many birds, deer and other wild animals from the forest for sake of diet. This result in decrease of their population.

## Deforestation

Deforestation is the result of expansion and urbanization which drastically affect the environment. We are exploiting the nature and environment beyond the safe limits. Due to it, wild animals are getting into villages and attacking humans.

## Excess Usage of Commodities

Purses, leather shoes, belts etc. are made of skin and hides of many animals. Cows, buffaloes, pigs are killed for their skins. Even the tiger, elephants are killed for their nails, bones, tusks etc. and for decorative items also. Thus, human's greed for more is causing them to be extinct in few years. Exploitation of raw materials of medicinal importance by commercial establishments like drug companies for trading has affected the medicinal flora.

## Habitat loss

Modernization patterns and trends create serious problems of habitat fragmentation for construction of road, dam, factories, residential flats etc. Developmental activities encroach into natural forest areas and deplete the plant wealth. To improve efficacy, hedges have been removed and this has reduced the amount of habitat available for wildlife. It has also increased soil erosion.

We often do things and use things that deplete the bio-heritage of earth and cause pollution which harm the entire living world. Each of us in our own little comfortable ways contribute daily to the destruction of ecology. It is time now to awaken in each one of us that respect and pay attention to our beloved mother earth deserves.

# Chapter 16

# Water Crisis

Water occupies a pivotal position among the geo resources of the planet earth. In fact earth is perhaps the only planet endowed with water in abundance. It is an essential part of our everyday life. The life owes its origin and sustenance on the presence and availability of water. Water is a bond between human beings and nature. It is one of the basic needs of human beings also. Life and water may be said to be two faces of the same coin. Though nearly 70% of world is covered by water only about 2.5% of it is fresh water. Even then 1% of fresh water is easily accessible for human consumption while remaining majority is tapped in form of ice caps and glaciers. Water is a renewable but finite resource.

Many of us, who are living in metro cities, enjoy an easy going and relaxed lifestyle with running tap water round the clock, swimming pools and decorative fountains. Enjoying these comforts, we remain unaware of the impact of these water extensive activities on our environment. Rapid urbanization and water pollution has broadened the supply and demand gap creating pressure on quality of water. When more people are competing for same amount of resource, it naturally intensifies the competition. But, water is the resource that seems to face the largest stress from exploitation of other natural resources including forests and minerals. Mining pollutes water and logging degrades water supply by denuding the forests that stabilize the watershed. Many of water systems that keep ecosystem thriving and feed a growing human population have become stressed. Rivers, lakes and aquifers are drying up or becoming too polluted to use and have dwindled into streams of garbage. More than half the world's wetlands have disappeared. When water run dry, we can not get enough water to drink, wash or feed crops and economic decline may occur. Besides, inadequate sanitation can lead to deadly diarrhea disease and other water borne diseases.

Fresh water is emerging today as a scarce resource being felt the world over, fuelled by population explosion, rapid urbanization, agricultural developments, erratic weather patterns brought by global warming, and poor waste water regulation. It was very difficult to imagine few decades before that we have to buy and pay for drinking water, so called "mineral water" bottles. Water has always been looked upon as a "free good" or a "nearly free good" because of its seemingly plentiful availability. But as one statesman puts it, "water has a price, whether it is paid or not". The use value of water was never determined, but it is time to give its due importance. More than 2000 million people would live under conditions of high water stress by year 2050 according to United Natural Environment Programme (UNEP) which warns that water could prove to be a limiting factor for development in number of regions in the world. About one fifth of world's population lacks access to safe drinking water and if the current rate of consumption patterns continues, two out of three persons on earth would live in water stressed conditions by 2030. Besides, climate change is expected to exacerbate the problem by diminishing the supply of water coming from rainfall and glaciers. It is predicted that in 21ˢᵗ century, there will be 'Water Wars'. Water promises to be the 21ˢᵗ century what oil has to the 20ᵗʰ century: the precious commodity that determines the wealth of nations. However, unlike oil, water has no substitute! Within a decade or two, water could overshadow oil as a precious resource commodity at the centre of conflict and peacemaking.

India is facing a water crisis that has implications not only for its billions people but for entire globe. The scarcity of water is a well known fact. In spite of higher average annual rainfall in India (46 inches) as compared to global average (32 inches), it does not have sufficient water. Most of rain water falling on earth surface flows away rapidly, leaving very little for recharge of ground water. As a result, most part of India is facing lack of water even for domestic use. The reality of water crisis can not be ignored. Ladakh is suffering from probably its worst water crisis. Exponential increase in tourists with decreasing glaciers (21% decrease in glacial area in western Himalaya) is casting a dark shadow on its present and future. The major source of water for drinking and irrigation is glacial melt water. According to Jigmet Takpa, Joint Secretary at the Ministry of Environment, Forests and Climate Change, if there is an abrupt rise in temperature , glaciers will be wiped out from the Himalayas, resulting in frequent flash flood and glacial lake outburst flood (GLOF) followed by a drought. The region is witnessing receding rainfall along with unexpected changes in its timing. Generally it receives maximum snowfall in November and January allowing the snow to accumulate and freeze. Now, it is witnessing a lot of snow in late January and February. With not enough of a glacier frozen before spring, they begin to recede a lot quicker resulting in water scarcity. Tackling the effects of receding glaciers and strong previous water resources, remarkable engineers from the region like Sonum Wangchuk and Chewang Norphel have constructed ice stupas. However, even these efforts are proving insufficient to current water crisis. The recent water crisis in Shimla, Himachal Pradesh is well documented.

The demand for water is already out stripping the supply in urban India. They got most of their supply from rivers and lakes. At one side, cities are facing severe water crisis and on other hand they are flooded with rainwater or faced drought. Today, the hunt for a water source and its sustenance is a big concern. In cities, majority of the population nowadays have become ground water dependent. As a result, ground water reserves are being tapped and overexploited resulting into decline in groundwater table and deterioration in quality. Extraction of groundwater is being done unplanned and uncontrolled, therefore, it has resulted in hydro-biological imbalance, deterioration in water quality and rise in energy requirement in pumping. Rainwater harvesting is an age old system of collecting rainwater for future use. But, systematic collection and recharging of groundwater is a recent development and is gaining importance as one of the most feasible and easy to implement remedy to restore the hydro-biological imbalance and to prevent the crisis.

Due to infra structure, the Government is unable to supply drinking water to its citizens. Water scarcity in India is predominately a man made problem. Extremely poor management, unclear laws, government corruption and industrial and human waste have caused this water supply crunch and the available water practically useless due to large quantity of pollution. Gross mismanagement in preserving rain water is much to blame for the present water crisis in the country of India. Therefore, if India makes significant changes in the way it thinks about water and manages its resources soon, it could ward off the crisis. India may avoid this dark future if people take action immediately, to start roof top rainwater harvest, treatment against industrial waste and regulate the utilization of ground water. Our ancestors were very wise to realize this fact and practices so as to conserve this commodity and because of this, we have inherited with enough of resource to survive. This Water-Wisdom at all levels of society ensured adequate availability of water for all which in turn the basis for development and prosperity. Let us expand this old wisdom for benefit of all our people. As the water crisis continues to become severe, the water management should be reformed with measures for revival to traditional system and wisdom.

The world now, is in a dire need of a 'Blue Revolution' to conserve and manage fresh water supplies in the face of growing demand from population growth, irrigated agriculture, industries and cities- just as the Green Revolution transformed agriculture in 1960's. A Blue Revolution will require coordinated responses to problems at local, national and international levels. Local communities when start managing freshwater resources efficiently, they will also be able to manage other natural resources better, improve sanitation, and reduce diseases. At national level, especially in water –short regions with dense populations, there is a need to adopt a watershed or river-basin management perspective. At international level, countries that share river basins can fashion workable policies to manage water resources more equitably. Besides, a general awareness is required towards the judicious use of water. A balanced water management approach, involving radical changes in policies, practices, performances and public behavior apart from changes in constitutional provisions, is required to bridge the gap between demand and supply. What we need is a change of our mindset and a bit of foresightedness.

Both with respect to quantity and quality, the country require a balanced policy consideration and certainly a more efficient management practices than what is prevailing currently. The demands for "sustainable water management" and increasing population require more potable water from a declining available potable water base. Proper water management is a critical component of sustainable development i.e. development that meets the needs of present and future generations both. In many industrialized cities, both surface and ground water are contaminated due to misuse of water resources as consequences of human activities. Besides, bad water management also can not be denied. Both now and in future, for good management of water resources and all other natural resources should be considered as common heritage of humanity. Water management should be as decentralized as possible – what local administration can do should not be done by regional, state or central administrations. The role of users is even greater in exploitation of groundwater. The need for a hydrological education of general public is more important for the effectiveness of water management. We can not afford to keep wasting and fouling our precious supplies of fresh water. It is time for widespread conservation measures, effective water management policies, and growing attention to assuring fresh water supplies and sanitation as a part of development.

Human beings could not save and conserve water and its resources, probably because of its availability in abundance. But this irresponsible attitude resulted in deterioration of water bodies with respect to quantity and quality both. Now situation has arrived when even a single drop of water matters and water conservation has become the need of the day. Clean drinking water is destined to become one of the rarest commodities soon if we will not be alert about the significance of storing, recycling and reusing water. To avoid a crisis, many countries must conserve water, pollute less, manage supply and demand, and slow population growth. Overuse and pollution of water also are taking a heavy toll on the natural environment and pose increasing risk for many species of life also. Whether water is used for agriculture, industry or municipalities, there is much room for conservation and better management. Efficient and effective water management is necessary. Now the time has come to grab hold of water where it falls. People need to be aware of the alarming water shortage that we are facing currently and the imminent danger of shortage in the future. When we should come together to not only save water whatever is left but to enhance it further so that we may transfer it to our next generation in adequate quantity. In the last, we can say that as people face the challenges of sustaining the world's water today and for future, religion continues to play an increasingly recognized ethical and practical role.

The prediction of Shri Sunderlal Bahuguna *"the acute scarcity of water may force the powerful nations to wage a new global war for the control of the depleted sources of water"* may also become true. So we should respect the United Nation's message *"water is a scarce resource to be managed and protected"*.

# Chapter-17

# Rainwater Management

Rainwater management is the process of managing each and every drop of rain water for productive use such as crop production, drinking water for people and livestock, industries and increasing groundwater recharge etc. without deteriorating land and groundwater depletion. India receives about 4000 billion $m^3$ water through rain during monsoon. As rainy season concedes with hot atmosphere and high evapotranspiration demand, up to 50% precipitation is lost through transpiration and another 30% runs off. In recent years, there has been an increasing trend of groundwater based irrigation system using electrical and diesel pumps. Therefore, in many parts of the country, ground water extraction has touched the exploitation level. Due to overexploitation leading to groundwater mining and limited recharge, the groundwater is getting depleted at an alarming rate.

The status of water in the environment is unique. It is a key component of environment and human being occupy the central position. Need of water for irrigation, for drinking and other domestic use by human and animals, for commercial use, for energy production etc. has brought into focus the fragility of environment and need to guard against it for sustainable growth. The depletion of forests has further aggravated the problem. Trends in water consumption indicate that demand for water for household and industrial uses in developing countries could double as a proportion of total water demand in the next 25 years. Therefore, rainwater management is need of the day.

Importance of Rainwater harvesting and management (RWHM) to solve many of world's problems, the promotion of RWHM is growing rapidly throughout the world. RWHM is believed to have a major role as a type of sustainable infrastructure by enabling developing countries to meet Millennium Development Goals and as a sustainable technology for developing countries.

## Need of Rainwater Harvesting

☆ To overcome the adequacy of surface water to meet our demands.

★ To arrest decline in groundwater level.

★ To enhance availability of ground water at specific place and utilize rainwater for sustainable development.

★ To improve groundwater quality by dilution.

★ To increase agricultural production.

★ To increase infiltration of rainwater in subsoil.

★ To improve ecology of area by increase in vegetation cover etc.

The term rainwater harvesting refers to direct collection of rain water falling on the roof or on to the ground without passing through the stage of surface runoff or land. It is a technique of collection and storage of rainwater at surface or in sub surface aquifer before it is lost as surface runoff. Rainwater harvesting may be one of the best methods available to recovering the natural hydro-biological cycle and enabling urban development to become sustainable. It is one of the most technically feasible methods particularly in rainfed areas. Runoff harvesting and recycling for crop production is an essential component of watershed management. The main base of rain water harvesting is-

★ Recurrent droughts are common due to climate of area.

★ Low fertility status of soil.

★ Low water retention capacity of soil.

★ For increasing productivity under rainfed conditions.

★ Drinking for livestock.

★ Ground water recharge.

★ To reduce flood flows and top soil loss.

★ To provide an excellent backup source of water for emergencies.

★ To supplement other sources of water supply such as ground water or municipal water connections.

Rain water harvesting can reduce the volume of storm water, thereby lessening the impact on erosion and decreasing the load on storm sewers. Decreasing storm water volume also helps storm water pollutants like pesticides, fertilizers and petroleum products out of rivers and groundwater. Conservation of rainwater in rural areas is possible through different rain water harvesting structures like ponds, percolation tanks, injector wells etc. depending upon the hydrological conditions of the particular area. Due to population explosion and scarcity of open areas in urban areas, rainwater is wasted. One possible and simple technique is to divert the water falling on rooftops to store underground spaces. Water harvesting forms an important component for the development and management of water resources for domestic, agricultural, municipal usages. Due to unreliable supplies from surface water, the dependency on underground water has reached its critical limits.

In-situ water harvesting play a key role in dry land to increase productivity and reduce runoff, soil loss and nutrient loss. It is the conservation of every drop of rain water falling on earth for reducing run off and soil loss, thereby enhancing crop production. As water and land management are inseparable and successful strategies on rainwater management are location specific and depend upon rainfall, slopes, soil types and texture etc. Basic principles for in situ rainwater management are:

☆ Opportunity for arresting rainfall at the site of occurrence by increasing infiltration rate by deep tillage, profile modification, vertical mulch and by keeping the soil surface rough to facilitate more infiltration.

☆ Reducing the velocity of runoff by creating barrier.

☆ Intercepting a long slope into several short slopes so as to maintain less runoff.

☆ Reducing evaporation from soil surface.

☆ Protection against damage due to excessive runoff.

☆ Efficient utilization of the conserved moisture by selecting suitable crops.

In arid and semi arid areas, low rainfall occurs with high runoff and poor soil moisture storage. This surface runoff from an area can be increased by ex-situ rainwater management by reducing the infiltration capacity of soil through vegetation management, cleaning, chopping surface vegetation and reducing soil permeability by using chemicals. Basic principles of ex-situ rainwater management are-

☆ Increase the time of infiltration into the soil by constructing barriers across the slope such as contour bunding, terracing and check dam etc. to recharge groundwater.

☆ Reduce the length of slope to reduce the velocity of flow.

☆ To prevent excess water and soil losses.

☆ Storing the runoff and its efficient recycling.

## Water Harvesting System- Our Age Old Tradition

India has an age old tradition of water harvesting or water management. The earliest reference of this practice in India is in mythology and dates back to Vedic Times. King Bhageerath is supposed to have arrested the water of river Ganga in its upper reaches and diverted it to a different cause by constructing a barrier across it or by selectively breaching the natural barrier created across it because of land slide. Recorded evidence of water harvesting is found in Harappan and pre- Harappan civilization dating back to 4000- 6000 B.C. The Indus valley cities had excellent system of water harvesting and drainage systems. Dholavira, grandest of cities of its time located in Kutch district of Gujarat, laid out on a slope between two storm water channels. A tank excavated at Sringaverpur near Allahabad, U.P. dates back to 1st century B.C. A fully brick lined tank sized 600x60x12 feet deep is a sophisticated water harvesting system that used the natural slope of land to store flood waters of River Ganga. Chola King, Karikala built the Kallanai across the River Cauvery to direct water for irrigation (it is still functional) while King Bhoja of Bhopal built the largest artificial lake in India. Through out India from Kashmir to Kanya Kumari and from east to Gujarat, several ingenious ways have been devised to catch and store rainwater for future use.

Most of the forts of India have elaborate system of water security as an integral part of design which is visible in case of forts of medieval Rajasthan. Rajasthan is basically part of arid and semi-arid climatic conditions with limited rains and

                                                    *Environment and We in 21ˢᵗ Century*

therefore, primary source of water supply was rain water. The arrangements made for water conservation in Rajasthan is very interesting. The various examples are- Jaigarh Fort near Amber, Alwar Fort, Mehrangarh Fort of Jodhpur, Fatehpur Sikri etc. which have rainwater harvest system. Builders of famous Chittor and Ranthambore Forts had vision of exploiting natural catchments in forts created by undulating hill tops.

This significantly reflects the wisdom of our forefathers who adopted water harvesting and its management an integral part of native culture and community life. This water-wisdom at all levels of society ensured adequate availability of water for all, which in turn formed the basis of overall development and prosperity.

## Is Rainwater Harvesting Successful?

Cherrapunji was always famous for highest rainfall in the world, it still does. This area is located in green hills of Meghalaya and despite of heavy rainfall, it experiences acute water shortage due to excessive deforestation, building construction and not using methods of water conservation. There has been excessive soil erosion and stretches of hillside devoid of trees can be easily seen. People have to walk long distances to collect water.

The story is quite different in the area around River Ruparel in Rajasthan where proper water conservation is practiced. This particular area does not receive even half of the rainfall received by Cherrapunji, but proper management and conservation have done that more water is available than in Cherrapunji. With the help of NGO (non government organization), women residing in that area were encouraged to build round ponds and dams to hold back rain water. Gradually water began coming back as proper methods of conservation and harvesting were meant. If we, human being put in an effort, the damage caused by us can be undone.

## Water Harvesting –cum- Artificial Groundwater Recharge:

Water harvesting including a wide range of techniques based on the source of water, required storage duration and intended use for urban and rural, are -

Water Harvesting Techniques

**Rooftop Rainwater Harvesting**

- Percolation Pit Method
- Percolation pit with bore well method
- Recharge Trench method
- Existing well method
- Existing bore well method
- Storage Tank method

**Surface Runoff Harvesting**

- Percolation
- Check dam
- Farm ponds
- Injector wells
- Spreading - Flooding
  - Basin
  - Channel
  - Pit and Shaft

Methods of Ground water recharge mainly are:

Urban Areas-

Rooftop rainwater/ Storm runoff harvesting through

1.  Recharge pit
2.  Recharge trench
3.  Tube well
4.  Recharge well

Rural Areas-

Rainwater harvesting through:

1.  Gully plug
2.  Contour Bund
3.  Percolation Tank
4.  Check dam/ Cement plug
5.  Recharge shaft
6.  Dugwell recharge
7.  Ground water dams

Rooftop rainwater harvesting in towns is an innovative method of harvesting rainwater and is becoming popular in many towns and cities. Rooftops of houses serve as excellent and economical form of collection centers for rainwater. If properly diverted and used for drinking purposes for human and animals, rising fruit plants and artificial recharge it will augment groundwater table to a sufficient extent. National and local government are creating new regulations and funding sources to encourage water harvesting. The Central Ground Water Authority has begun awareness campaigns on popularising rooftop rainwater harvesting across the country. Rooftop rainwater harvesting system is now mandatory for new buildings in 18 of India's 28 states and 04 of its seven federally administered Union Territories, according to India's Press Information Bureau. When builders and architects are designing a new building, apartments or home, it is essential and important that they think of implementing rainwater harvesting methods. Karnataka state is also considering a water bill rebate for citizens who install such system in their houses. The Ministry of Environments and Forests has also brought out some measures on rainwater harvesting such as -

☆  All buildings constructed in urban areas should have rooftop rainwater harvesting system.

☆  Rainwater collected through storm water drains should be utilized to clear water disposed drains.

☆  In rural areas, rainwater harvesting should be through percolation tanks, storage tanks etc.

☆  Ground water aquifer recharge structures should be constructed.

It can not be denied that sustaining and recharge of groundwater along with judicious use of limited freshwater resource is need of the hour. If current condition continues and sufficient measures are not taken up immediately, we should be ready to face a crisis which will be detrimental to the survival of mankind. Efficient management of water resources and awareness about judicious use of water resource along with measures of recharging and maintaining the quality of water has to be adopted immediately. A planned attitude is required in order to fully utilize the potential of rainwater to meet out our demands. It will be able to reduce reliability on natural resource for groundwater. Therefore, if rainwater is conserved and utilized through rainwater harvesting technique, it may prove to be an effective tool of replenishing groundwater resource. In the long run, there will be energy savings, water savings and resource savings.

# Chapter-18

# Religious Belief and Environmental Protection

Ever since the birth of human beings on the planet earth, man has been completely dependent on natural resources around him for his survival and basic needs. Since the very beginning, man has used plants for food, fodder, fuel, fiber, spices, oil, timber, and medicine. If plants can express themselves, they will have the same opinion for man as man has for the parasite like flea, germs and bacteria living on him. But, man's ambition for limitless comfort has led him towards the exploitation of nature's wealth and resources indiscriminately and rapidly. Man's greed for resources and his desires to conquer nature has placed him in front of the environment.

Our relationship with nature or environment is best reflected in the form of religious belief and its practices. Environmental science and ecology are disciplines of modern science but their origin can be seen long back in Vedic and ancient Sanskrit literature. Our ancient forefathers perceived God's presence around them through nature. They felt that they must live in harmony with all of God's creation, respect and revere all of nature and divine forces. They identified the divine forces as air, water, earth, fire and sky. Vedas have depicted the close relationship between man, nature and religion. Many hymns were written in the Vedas praising the elements, rivers, forests, animals and sun and considering worthy of worship. Religion was probably used in ancient India as a tool to protect nature and natural resources and several instances of worshipping the trees have reported from different periods of the country. Yajurveda emphasized that the relationship with nature and animals should not be that to misuse the resources and therefore, there was a check on excess utilization of resources. Similarly, Rig-Veda is celebration of nature. It highlighted the potentialities of nature in controlling the climate increasing fertility. Atharvaveda considered trees as abode of various Gods and Goddesses. Also the Skand Purana states Tulsi and Bilva Tree as abode of Goddess Laxmi and Parvati respectively. Banyan tree is considered to be very sacred and worshipped as the abode of Trimurti's reside.

*" Parvati Bilva Vrikshastham Laxmi Cha Tulsi Gatam."*

In general practice, there are many flora and fauna which are sacred among the followers of Vedas because they are directly or indirectly related with different Gods. Animals were revered too. Kamadhenu was the wish-fulfilling cow, whose offspring are all the cattle on earth. Krishna even lifts Mount Govardhana to save cattle from Indra's wrath, a recurring theme in Indian art. The festival of Nagpanchmi –snake worship is celebrated as a thanks giving after the harvest season. But the greatest honour given to animals was their elevation as the vehicles of the Gods/ Goddesses (Table 1.1). Similarly, there are many sacred plants and Hindus worship them regularly, for example- *Oscimum sanctum* (Tulsi), *Eleocarpus sphaericus* (Rudraksha), *Ficus bengalensis* (Bar), *Ficus religiosa* (Pipal), *Prosopis cineraria*(Shami) etc. Some of these herbs were believed to cure every human disease from head to toe. A few of these plants were so common in our daily life and their therapeutic value was considered so great that they became a part of religion and worship by Hindus. In Rig-Veda, 'Soma' (*Ficus benjamina*) is mentioned as king of plants. Sometimes threads are tied around certain trees to seek a boon. *Nyctanthus arbor tristis* ( Parijat / Harsingar) is another divine tree believed to be introduced by Lord Krishna from the Heaven. There are hundreds of medicinal plants which are in use from the Vedic periods till today. Oushadhi Sukta of Rig-Veda addresses to plants and vegetation as mother. Several Indian trees and shrubs were regarded as sacred because of their medicinal / aesthetic / natural qualities as well as their proximity to a particular deity. Tree symbolized various attributes of Gods to Rig-Veda Seers. Association of trees, plants and animals with various Gods and Goddesses at that time testified reverential attitudes of people towards the environment (Table.1.2). Besides, indigenous people worship various God and Goddesses in the vicinity of their abodes. These places of worship are popularly known as sacred groves. This religious faith and belief of these people acts as binding force for the local community. Sacred groves are patches of forest lands, untouched by development agenda, where gods or goddesses are worshipped. Sacred groves balance the ecosystem in forest cover. Certain rare, valuable and medicinal flora and endangered animals are protected from extinction. These beliefs are eco-friendly in nature and have a scientific interpretation in ethnobotanical terms.

Totemism refers to the religious faith of tribal communities in animate objects which they consider as the embodiment of divine power. Tribal inhabit geographic locations untouched by human interference in the forests and worship the Mother Nature. Having religious belief in nature's objects especially plants and trees, they act as agents of protection and conservation of the environment.

**Table. 1.1** – A General List of Animals Favoured By God/Goddess

| *Animal* | *Favoured by God/Goddess* |
|---|---|
| Owl | Goddess of wealth (Laxmi) |
| Bull | Lord of animal (Pashupatinath),Lord Shiva |
| Lion and Tiger | Goddess of power (Durga , Kali etc,) |
| Horse | Lord Sun |
| Serpent | Lord Shiva |
| Monkey | Lord Hanuman |
| Dog | Lord Bhairab |
| Rat | Lord Ganesh |
| Swan | Goddess of knowledge( Saraswati) |
| Eagle(Garun) | Lord Vishnu |
| Cow | Lord Krishna |

**Table 1.2** - A general List of Plants/Trees favoured by God /Goddesses

| *Tree/Plant* | *Favoured by God /Goddesses* |
| --- | --- |
| Neem | Sitla, Mansa |
| Bel | Parvati |
| Tulsi | Lakshmi |
| Vishnubela | Shiva |
| Ashok | Buddha, Indra |
| Kadamba | Krishna |
| Mango | Lakshmi |

Mythology also has been useful in cultivating certain plants, for example- Tulsi, a highly valued medicinal plant is grown in every household and ritually watered daily even today. This is clearly a case of religion and culture being used to protect, conserve and / or produce resources from human sustenance. The students of Vedic period lived in a natural environment (Gurukul) and protected trees and worshipped them as Vriksha Devta (Tree God), the forest covers as Van Devta (Forest God) and the rivers as sources of delicious life-giving water. The ancient people cared for wildlife too. Terms and titles such as Nag Devta (Snake God), Kamdhenu (the cow that fulfills your desires) and Kalpavriksha (the wish –fulfilling tree) symbolized the benefits that accrued to human beings from nature and their respect for wildlife. The ecological imbalance caused by the criminal acts of the so called 'civilized man' has resulted in a disastrous threat, not only to the human survival, but also to life as a whole on the earth.

India is a land of rites and rituals. Almost all major religions of the world are represented in India. Popular belief states that Lord Krishna played his flute under the tree of Kadamb. Likewise the worship of 'Putrranjiva' tree results in a long and prosperous life of male off-springs. Besides Hinduism, all other religions also realized the proximity of mankind with nature. For example-  Basic tents of Buddhism are simplicity and non-violence. Both these principles of Buddhism are of great importance in the conservation and protection of natural environment. Gautama Buddha attained enlightenment under the branches of Ashok tree. Similarly King Ashok wanted the non-violence to be the cultural heritage of the people. Therefore, punishment was prescribed for killing animals. Jainism and Islam also are also based on the principle which is in close harmony with nature and propagates respect for nature's creation and their protection.  There is a close link between Christianity and environment and a thrust for sustainable development. Sikh religion is comparatively of recent origin. Guru Granth Sahib Ji also emphasizes that human being are composed of five basic elements of nature. Thus close relationship between nature and mankind has been recognized. Therefore, the cultural and religious heritage of India shows deep concerns for the protection and preservation of the environment. All these religions realized the proximity of mankind with nature.

History of recent past tells us it was not only Kings or rulers of India who showed concern for nature but even common men were deeply involved in strong environmental practices. For instance, Bishnois of Rajasthan involved themselves in the respect for all living things and their protection. It is believed that 360 of them mostly women and children sacrificed their lives by hugging the trees to save them from axe men of Maharaja of Jodhpur in 1730. Chipko Movements is one of the most successful conservation movements in India by womenfolk of Garhwal in Uttarakhand.

## Plants and Planets

The stars and planets seen in the sky must have bewildered man since the very beginning. An association of stars and planets with deities has also been a mythological belief since time immemorial. The stars and planets act merely as pointers to the auspicious and inauspicious moments in the life of human beings. Navagraha is the worship of the nine planets of solar system and some plant species are worshipped as the symbolic representation of these planets (Table 1.3). These plant species are believed to have resemblance with deities. The planets also affect plants and animals. The Sunflower capitulum moves with the movement of Sun, opening and closing its flower in a diurnal cycle. It is the example of the relation between planetary movement and biological activity. The rulership of planet over plants can be shown by some examples- Mercury governs memory and *Bacopa manniera* ( Bhrahmi) governed by mercury acts as a medicine for memory. The fruits of *Juglanas regia* (walnut) appear as cicumvolutions of the brain. Mercury governs the brain and its activity is governed by Sun. Both these planets govern the walnut. The plants which are governed by a particular planet have the property to cure the diseases caused by the respective planet.

Another type of analogy between planets and herbs is that some parts of herbs used as a substitute for a particular gem are assigned to a planet (Table1.4). Besides, plants were classified into nine groups to match the nine planets for offering into the 'Havan' as 'Samidha' shows the classification as given in Garuda Purana.

*Table. 1.3* - List of plants associated with nine planets and Samidha

| Name of Planet in English | Name of Planet in Hindi | Tree/Plant |
|---|---|---|
| Mercury | Budha | *Achyranthus aspara* (Apamarg/Latjira) |
| Venus | Sukra | *Ficus glomerata*(Audumber/ Gular) |
| Mars | Mangal | *Acacia catechu*(Khair) |
| Jupiter | Guru | *Ficus religiosa*(Pipal), Musa sps.(Kela) |
| Saturn | Shani | *Prosopsis cineraria*(Shami) |
| Sun | Surya | *Calotropis procera* (Aak/Madar) |
| Chandra | Som | *Butea monosperma*(Palash) |
| Neptune | Rahu | *Cynadon dectylon*(Durva) |
| Pluto | Ketu | *Desmostychabipinnata*(Kusha) |

**Table. 1.4** - Plants, planets and Gems

| Botanical name of plant | Common name | Planet | Gem |
|---|---|---|---|
| *Aegale marmelos* | Bael | Sun | Ruby |
| *Mimusops hexadra* | Khirni | Moon | pearl |
| *Opuntia dillenii*<br>*Hemidasmus indicus* | Nagfani<br>Indian<br>sarsaparilla | Mars<br>Mars | Coral |
| *Curcuma longa*<br>*Musa sapientum* | Haldi<br>Kela | Jupiter<br>Jupiter | Yellow sapphire |
| *Punica granatum*<br>*Ricinus communis* | Anar<br>Arandi | Venus<br>Venus | Diamond<br>Zircon |
| *Lettsomia nervosa* | Argyreia | Mercury | Emrald |
| *Pterocarpus santolinus* | Lal chandan | Saturn | Blue sapphire |
| *Santalum album* | Chandan | Rahu | Gomed |
| *Withania somnifera* | Ashwagandha | Ketu | Cat's eye |

Vedic people desired to live a life of hundred years and this wish can be fulfilled only when environment will be unpolluted, clean and peaceful. An India's relation with nature differs from that of western men. In the West, man has separated himself from nature; he believes and used it to serve his own purpose. Hindu unites himself with nature. From nature he came, to nature he returns as ashes. Modern scientists also feel proud of our ancestors for their knowledge and views about the environment. Vedic message is clear that environment belongs to all living beings, so it needs protection by all, for welfare of all. Indian religious literature is replete with ideas of forest conservation, utilization and regeneration. Our ancestors had left various religious beliefs for us towards nature and it was a very constructive device for conservation of environment and biodiversity not only during their time but at present also. Therefore, main reason for associating plants and animals with religious rites was probably for conservation of biodiversity.

Environmental protection and sustainable development are concerns of such nature that permeate the entire humanity and wisdom of protecting nature. In recent years, we seem to have lost touch with our glorious tradition and wisdom to protect nature. In effect, environmental conservation is embedded in the Hindu consciousness, being a good Hindu means being a champion of the environment. Mahatma Gandhi has delivered inspirational message for environmental movement. He has said *"The country's development has to be in harmony with nature. The earth has resources to meet everybody's needs but not anybody's greed"*. On the other hand, advancement in science and technology followed by development and education of masses in the current scenario, young generation  consider these religious faith as orthodox and conservative. The strength of religious faith as binding force between man and nature is slowly fading. But, some religious practices interpreted as conservation techniques are backed by logical/ scientific reasoning. The legend 'Old is gold' is more appropriate in present concept. Our ancestors had a faith that God dwells in plants and trees, so they must have to offer worship to them. Therefore, they have left various religious beliefs for us towards nature and it was a very constructive device for conservation of environment and biodiversity not only during their time but at present also.

Therefore, it is evident that Indian rituals have been contributing a lot in protection and conservation of environment in India since Vedic Period. Hence, it is need of hour to go for these rituals and traditions in order to keep up the balance of nature. The ancient Indian message is that the environment belongs to all living beings, so it needs protection and conservation by all, for welfare of all.

# Chapter-19

# Women and Environmental Protection

Since time immemorial, women have long been allied with nature. Earth is surmised to be feminine in nature and it has often been metaphorically termed as 'Mother Earth.' Woman symbolizes 'Prakriti' meaning 'Nature' in Indian Philosophy. She creates and nurtures the creation to bloom. She signifies 'Shakti'- the power that derives the system. Women have been generally submissive as has been nature. Vedic script considered women as Goddesses of environment and her revered position as the conceiver of ecology was considered as extraterrestrial. But, the status of Indian women has eroded significantly and her role was sidelined in environment. Nevertheless 20th century saw reawakening and upswing to enhance educational and social reforms. Women always play a vital role in environmental management and governance. Since time immemorial, women are traditionally involved in protecting and conserving their natural resources. They have more sympathetic and positive attitude towards conservation. Since they take a pivoting position in the family, day to day work of tending their gardens and feeding their families in developing world, women have become unrecognized heroines of environmental conservation and sustainable development. From global perspective, if we talk about the natural resource management, it is the women only who comes in forefront of protection and preservation of resources and ecosystems.

There is high degree of relationship between women and nature as revealed in historical analysis of functional relationship. Women particularly those living in rural or mountain areas have special relationship with the environment. They are more close to nature than men and this very close relationship makes them perfect managers of an Eco-system. At one level, Nature is symbolized as the embodiment of feminine principle and at other end, it is nurtured to provide life and sustenance in hand of women. This special linkage of women and environment is critically damaged by modern growth oriented economic development. The growth in modern Science and Technology resulted in violence both against women and environment. This led to marginalization of concern for crucial role that they play in survival and sustenance of life.

Women's environmental concerns were first highlighted in 1975, the International Women's Year which was declared in honour of 25th anniversary of United Nation Commission on the status of women. Women have a vital role in environmental management and development. Their full participation is, therefore, essential for achieving sustainable development. Besides conserving nature, they also play pivotal role in management of domestic economy. The interaction of women with the environment as farmers, food producers and household managers has a direct impact on the well beings of the nature. One feminist ecologist puts that women have a truly productive relationship with nature because "women not only collected and consumed what grew in nature but they made things grow". Women and nature both work quietly and create wealth which is invisible but valuable because they ensure both stability and sustainability. Linkage between women and natural environment is profound and evident among rural women of third world like India. More than 70% of Indian population is rural based. Nature plays crucial role in meeting daily survival needs of vast majority of rural households.

## Women-Nature Connection

There is manifold connection of women to that with nature. These connections provide sometimes competing, sometimes mutually complementary or supportive of two dominations of women and nature.

*Women and Land :* Women produce most world's food and are engaged in all aspects of agriculture from planting, weeding and harvesting to livestock management and marketing if they have equal access to resources such as land, training efforts, techniques and credit etc.

*Women and Water :* Since women spend most of their day time in collecting water to meet household and domestic livestock requirements, therefore, they are linked to active resources.

*Women and Forest :* Women and girls are responsible for food and fodder needs of household. As gatherers of fuel wood, animal dung or charcoal for fuel, cutting fodder grasses, and other forest produce etc., they are linked to forests also.

This symbiotic relationship between women and environment was consistently maintained before the advent of modern science and technology and modern industry. The interaction of women with natural resources has important implications for biodiversity, desertification and water management. The life of mountain woman is so much intertwined with environment that whole Eco-system revolves around her and she can't even think of her survival without it. For her, forest is her mother's home as she is entirely dependent on the forest to meet her daily needs.

In present era, we are witnessing a remarkable convergence of policy objectives between themes of sustainable environmental conservation and advancement of women. Women's role and concerns in environment conservation remain poorly and incompletely acknowledged and rarely getting the credit for nurturing these life support system.

## Women and Eco–System Management

Women have always been the principal conserver of biodiversity and have vital role in management of sustainable Eco-system. Since time immemorial, women are traditionally involved in protecting and conserving their natural resources. Vandana Shiva and Bina Agarwal both asserted that women should be the central actors in environmental management because of their closeness to resource base to their daily survival tasks. They believed that being the privileged caretaker of the environment is the key to women empowerment.

With their extra ordinary skills and traditional knowledge, women have proved how land, water, forest and other natural resources can be used and managed. Through their practical experience and managerial skill, they have acquired knowledge of various types of plants, grasses, medicinal plants, kind of fuel wood and various species of fodder plants. They know better than any scientist that what grasses, herbs, shrubs and trees are best for them and should be planted to maintain a balanced eco-system and well being of families and communities. They are perfect in making an optimum use and conservation of natural resources.

Priorities of both men and women differ in forest usage. Women without harming the regenerative capacity of forests, extract fodder and minor forest produce. Men on the other hand use forest for sale, for house construction and for agricultural implements. In selecting trees, men choose commercially viable variety while women opt for trees which give fuel and fodder. But their traditional knowledge is being threatened by biodiversity loss and biopiracy (by pharmaceutical companies seeking to control use of medicinal plants).

## Women in Environmental Protection: Indian Scenario

Women play creative role in conserving, retrieving and regenerating the natural environment and its resources by protesting against environmental degradation, by opposing the unsustainable developmental initiatives of market forces and state apparatus and also by propagating environment friendly economic practices. They have been raising their collective voices in this regard at the global, national and local levels. Past few decades have witnessed an enormous interest in both the women's movement and environmental movement.

At the global level, role played by women delegates at Earth Summit 1992 at Rio-de-Janeiro is noteworthy. Similar voices have also been raised in various other International conferences stressing the relationship of nature and women.

As an instance in Indian scenario, individual women with her associated organizations has been making efforts to establish the symbiosis between communities, women and natural resources. Ela Bhatt of 'SEWA', Medha Patekar of 'Narmada Bachao Andolan', Vandana Shiva and Kamla Choudhary of National Wasteland Development Board etc. are they who raised their muted voice of Indian women at national levels. But the significant work done by rural hill women against destruction of forest in Himalayas goes to 'Chipco Andolan' in Garhwal region of Uttarakhand. Women of this region have successfully embraced the trees to save them from on slaught of forest contractors supported by state and its machinery between 1972-78.

After Chipco movement, rural local women in many villages of central Himalayas Mahila Mangal Dals (village level women group) have taken initiatives to regenerate the forest by planting trees and protecting them. The initiatives taken by local, rural, poor women from villages of Bihar, Madhya Pradesh, Orissa, Rajasthan, Gujarat, Andhra Pradesh, Himachal Pradesh etc. showed remarkable results in regenerating their degraded forests in last three decades. In various movements against developmental projects like- Sardar Sarovar project in Madhya Pradesh, Maharashtra and Gujarat, Tehri Dam project in Uttarakhand, Koel Karo Dam in Bihar, Subarnarekha project in Orissa etc. are considered as environmental movements involving women. They participated in these movements because their interest was worst affected with displacement of these projects. Similar is the case in West Bengal where Mamta Banerjee protested against the establishment of Ratan Tata's Nano car manufacturing unit to save the agricultural land of villagers. Many self help groups of women in India ( Periyar, Kerala) have been formed for patrolling the forest. The presence of a group of 6-8 women patrolling forest during daytime not only discourages illegal entry but also controls biomass extraction. Periyar is emerging as a role model of women empowerment for biodiversity conservation. Acting against the advice of Forestry Research Institute, women planted oak trees in deforested region so providing a new basis for water, fodder and fertilizers.

Mrs. Maneka Gandhi is another woman to start movement for conserving and protecting the animals of the nation from illegal and brutal killing. Though the women participants are not environment conscious, their complete dependency on forest resources for survival makes them active proponents of environmental conservation and regeneration of natural ecosystem.

Even today, they perform duties of such as seed selection, multiplication and conservation. Klaus Toepfer, Executive Director of United Nations Environment Programme (UNEP) said *"It is clear that women have a central role as custodians of local and indigenous knowledge and as conservators of the natural world"*.

A report from UNEP describes that role of women and their 'know-how' is often undervalued and ignored. Indeed all the women are treated as second grade citizen with less rights and reduced status in respect of men. There is need to give special attention to making them an equal partner in policy framing or movements to empower women. As the world moves forward at a phenomenal speed with science and technology, there is a growing feeling that biotechnology empowerment of women is absolutely essential for progress. With this in view, Department of Biotechnology has initiated programme since 1988 to empower women and rural population by imparting skills for additional income generation. As women have deep relationship with all components of eco-system, they should be given opportunity to participate in Village Eco-system Planning Training. Women's knowledge, skills, their traditional values and experiences must be recognized.

It is established without doubt that women play a major role in conservation and restoration of environment. The symbiotic nurturing relationship exists between women and their natural environment. Besides, social empowerment of women and environment is also closely linked. Economics and environment are compatible for poor women. Environmental degradation is related not only to biosphere but to social sphere as well.

Though Government of India is working towards an environmentally sound and sustainable quality of life, the problems, challenges and issues are multifaceted. However, women in India are playing a crucial role in protection and conservation of environment. Thus, today neither policy makers nor academicians can ignore the relationship of women and environment and their positive attitude towards wildlife conservation.

# Chapter-20

# Environmental Protection through Education

Environment is the sum total of land, water, air, man, flora and fauna and their interrelationship among themselves. Environment which human beings have all the times lived in and depend upon for survival is dynamic, variable and changing due to the forces within itself as well as activities of man. The human induced changes have damaged the physical environment to the extent that survival of man and other species has become a great challenge. The burgeoning population, unplanned urbanization, deforestation, industrialization, poor water quality, and unsustainable exploitation of natural resources etc. are seriously threatening the delicate balance between human beings and their environment. Man has continuously tampered with nature. As a result, a threat to his life has been increased due to lack of healthy air, water and imbalance in biosphere. Our earth is our home. So if we want to protect our home, we should protect our environment from harmful effects of human activity. Therefore, it is dire need of the day to find out new solutions to such problems of various pollutions. All these should be a part of environmental education.

Environmental education is study of environment- through the environment, about the environment and for the environment. It deals with the need to protect environment because global warming, pollution and many other issues are ruining our environment badly. Environment itself is a teacher to teach us along with other living organisms how to adapt and adjust in adverse conditions. The basic aim of environmental education is to create awareness among people, impart knowledge, inculcate attitude of concern and provide necessary skill to handle the environmental issues and challenges. Awareness means having latest knowledge of possible danger of a thing. By environmental awareness, we mean having update knowledge of possible threats of environmental degradation. It makes people conscious of physical, social and aesthetic value of environment. All these are interdependent, interrelated and necessary for survival of humanity. Environmental education gained importance at global level after the Stockholm Conference on Human Environment organized by UNESCO in 1972. After this conference, UNESCO launched the International Environment Education Programme (IEEP).

## Why Environmental Education is Need of the Day?

During last decade, environmental issues have been receiving increasing attention in all spheres of life because environment is degrading at a much faster rate than our imagination. Most of this mess is the outcome of human activity. Thus, what so ever wrong has been done by us must be rectified by us only. To protect, conserve and manage the environment, it is of vital importance to have a sound environmental education. It is a way to teach society, people and individual, how to utilize the present resources optimally. Such an approach to education becomes a medium and process of creating awareness about man's relation with his natural as well as social and man made environment.

## How to Impart Environmental Education?

The environmental education may be imparted through formal and / or non-formal mode. The formal education includes the Primary, Secondary and environmental studies in Higher Education also. The non-formal education covers the forest dwellers, poor villagers and tribal people who have no formal schooling and are illiterate. They should be taught to conserve environment, to be eco-friendly and co- existent. The importance to worship the trees, rivers, nature and earth should be logically explained to them. The five elements of nature are part of our life support system. To educate them, radio, Television and Audio-visual aids can be used.

To make environmental education more effective, new methods of teaching may be introduced from primary to college levels. Chart, poster, models, debate, power point presentations, group discussions, plays, exhibition, projects, and slogan writings etc. are some of the ways to develop awareness in students about their responsibilities regarding environmental protection. Institutions should arrange the Nukkad Natak (Street plays) to create awareness in general public about environmental issues and problems. Educational excursion visits can be organized to study a chosen natural area, wildlife, flora and fauna. Zoo and botanical garden visits will help children to get knowledge about environment and their protection. The time has come when this subject has to be a compulsory subject at all levels. However, to qualify the environmental studies paper is compulsory for graduation in some universities. Universities in India focus on teaching, research and training. In more than 20 universities, different colleges and Institutes, courses in Environmental Engineering, Conservation and management, environmental health and social science are taught. Supreme Court has also mandated that environmental education must be imparted at all levels including Higher Education in formal system. Even Bar Council of India has introduced 'Environmental Law' as a compulsory paper for legal education at the graduate level.

Values play important role within environmental education. All human values are linked to satisfying our biological needs. Since all we have comes from what the earth provides, there is an obvious link between our environment and our values. The environmental values we have come down to matter of making choices about the environment. Should I use CFL instead of incandescent bulb? Can I walk instead of drive? Do I require buying a new item, repairing the old one or doing without? Should I curtail our shower time to use less water? Answer to all these questions is actually personal and gives considerations to convenience, economics and social acceptance.

Women can also play a significant role in environmental management and protection. They are good educator for their children and family members shaping up and modifying their mindset. Many self help groups of women in India (Periyar, Kerala) have been formed for patrolling the forest and these self help group teaches various aspects for maintaining the environmental issues in school and colleges. Individual women with her associated organizations have been making efforts to establish the symbiosis between communities, women and natural resources. Ela Bhatt of 'SEWA', Medha Patekar of 'Narmada Bachao Andolan', Vandana Shiva and Kamla Choudhary of National Wasteland Development Board etc. are some names of Indian women who raised their voices at national levels. But the significant work done by rural hill women against destruction of forest in Himalayas goes to 'Chipco Andolan' in Garhwal region of Uttarakhand. Women of this region have successfully embraced the trees to save them from onslaught of forest contractors supported by state and its machinery between 1972-78.

## Why Environmental Education is Mandatory?

Environmental education must target the routine and how simple changes in a daily life can make major changes to environment. Because of change in societal setup, today's children are busy playing indoor games, electronic gadgets, watching television, playing video games or surfing the internet etc., they have no interest as well as time to go outside for playing around the natural environment. After becoming grownup as adults, they remain least bothered about the environmental protection. Hence, integrating environmental education into curriculum is best option to relate students with the nature right from their childhood. Environmental education can also encourage people to reduce their negative impact on environment through more efficient use of water and energy, particularly in areas where the resources are in scarcity. For example- Some educated farmers use rainwater harvesting and alternative techniques for irrigation. In urban areas also when most of the adults completed their primary education, purification of water increased by 9% while this figure changed to 22% where most of the adults completed their secondary education. Environmental education emphasizes co-operative learning that means working in teams, critical thinking, hand-on activities and action strategies. Therefore, students studying environmental education develop some leadership skills like-

- ✰ Working in teams.
- ✰ Listening and accepting diverse opinions.
- ✰ Solving real world problems.

☆	Promoting activities that have positive impact.
☆	Connecting themselves with the local communities.

Public awareness about different laws for wildlife and biodiversity conservation, conservation of natural resources, pollution control, emission of green house gases and their implementation and making of new policies are the goal of environmental education for environmental protection. It is a multidisciplinary subject where different aspects have a holistic approach. We all know that it is very difficult to modify practices and attitudes overnight. It is time to join hands for positive change. Education is not a magic stick or silver bullet, it must be supported with political leadership. The responsibility for protecting environment should be shared between individual and the government. If everyone change their habits, the day when people live in a both materially and ecologically balanced environment, is not so far. Therefore, environment should be protected for a better life in future.

Chapter-21

# Environmental Protection and the Judiciary

From the dawn of 21$^{st}$ century, the environmental issues have emerged as a major concern for the well being of the people. The conception of ecological protection and conservation is not modern. Ancient Indian texts give prominence that it is 'Dharma' of each individual in the society to protect nature. But, man himself is the maker and mentor of his own life. The environmental degradation occurring around the globe is somehow due to anthropogenic activities. Besides, day to day innovation and advancement of technology, apart from development additionally adds the risk to human life. Therefore, there arises an intense and dire need of the law to keep pace with the need of the society along with individuals. So, the environmental protection has become a matter of global concern. India has enacted various laws to deal with environmental issues. The law can be defined as a body of rules or norms governing, promoting, regulating and prohibiting human behaviour. The changing rate of environment is so fast that in order to keep the law on the same rate , either laws should be amended to cope up with new challenges or it should has some new directions under judicial interpretation. Environmental laws are easier to draft but difficult to implement due to conflicts between genuine needs and unwanted greed. Laws are made at three levels- International, National and Local/regional levels. Indian laws are enacted by Parliament in the centre or Legislative assemblies in the states (called- the Acts) or by governor of the state.

United Nations conference on Human Environment held at Stockholm (Stockholm Declaration, 1972) is the first document to declare the right to a healthy environment with great emphasis on protecting both species and their habitats. The declaration states *"Man has the fundamental right to freedom, equality and adequate conditions of life in an environment of a quality that permits a life of dignity and well being"*. The need to preserve the environment was not placed in opposition to economic development but essential for protection of human rights. India has a highly developed judicial system with the Supreme Court having powers to make any order for doing complete justice in any cause and a mandate in the constitution, to all authorities. Therefore, Supreme Court in 1972 worked as a catalyst in development of environmental jurisprudence in India. The aim was to protect and improve the natural environment including forest, lakes, rivers and wildlife and to develop compassion for living organisms. After Stockholm declaration, protection of environmental legislation started in India under the obligation of Supreme Court in 1978 (Wildlife Act 1972; Water Act 1974; Forest Conservation Act 1980; Air Act 1981; and finally Environmental Protection Act 1986). An independent department of Environment was established by Government of India in 1980 to impart environmental awareness but it has no power to prosecute the defaulters. In these circumstances, Indian judiciary needs special attention. Indian judiciary has expanded the existing legal provisions to address the problems of environmental degradation. It also gave various directions, orders and guidelines to check the environmental pollution.

At the turn of the present century, when world started to experience the negative impacts of industrial revolution, "Right to live in Healthy Environment" gained momentum.  The right to live in clean and healthy environment is not a recent invention of higher judiciary in India. It was only late eighties and thereafter, various High Courts and Supreme Court of India have designated this right as a fundamental right. In India, High Courts have substantially enriched environmental jurisprudence being the first to come up with direct and specific pronouncement on citizens' fundamental rights to pollution free environment. Therefore, Andhra Pradesh High Court ruled in 1987 that Nature's gifts without which life can not be enjoyed. Karnataka High Court on the same pattern pointed out that "Entitlement to clean environment is one of the recognized human rights". Kerala High Court again said that Right to Sweet Water and Right to Free Air are attributed to Right to Life. The High Courts and Supreme Court have shown keen interest in protection of environment. Thus, judiciary in India has played a significant role in interpreting the laws in such a way which not only aided in protecting environment but also in promoting sustainable development. The 42nd Amendment to Indian Constitution and insertion of Article 48-A and 51-A (g) marked the beginning of environmental protection in India. Environmental protection is not only a duty but also a right under Article 21 of Indian Constitution. Under Article 32 one can file Public Interest Litigation (PIL) in case of infringement of his rights given in the constitution.

The laws made by legislature for environmental protection are:

*For Control of Pollution:* Indian Penal Code contains provision for prevention of pollution  under sec. 277 ( about polluting water resources), 278(about making atmosphere noxious to health), 428 and 429( for protecting animals from all forms of violence against them) and many others.

The preamble of Air (Prevention and Control of pollution) Act 1981states: "An Act to provide for prevention, control and abatement of air pollution". Similarly, The Preamble of Water (Prevention and Control of pollution) Act 1947 states: "An Act to provide for the prevention and Control of Water Pollution and maintaining or restoring of wholesomeness of water". Both these acts contain provisions regarding the constitution of board of control of pollution, its jurisdiction and penalty for pollution of environment.

*For Protection of Forest and Wildlife:* The overexploitation and loss of habitat constitute a threat to the rich forest cover and biological diversity available in the country. So, The Indian Forest Act 1927 has been enacted for the protection of forests and to consolidate laws relating to forests. The Wildlife Protection Act 1972 which provides protection of wild animals, birds and plants. It has been passed on request of states under article 252 of Indian constitution.

*Biological Diversity Act 2002:* There is provision of punishment under this act. The offences are non-sailable with 5 years imprisonment and 10 lakh fine.

*Hazardous Wastes (management and handling) Rule 1989:* Under this act, proper handling and careful disposal of hazardous wastes like mercury, arsenic, paint, dye etc. is explained. The Hazardous Wastes (Management and Handling) Rule has instituted in 1989 and amended in 2008.

The Environment (protection) Bill was introduced in the parliament after the Stockholm Conference to provide protection and improvement of environment and for matters connected therewith.

## Judicial Approach towards Environmental Cases

### Pro-Green and Pro-Environmental Approach

Since the intervention of judiciary in environmental litigation from 1980-2000, it is found that there had been a common thread of pro-environmental approach followed by the Supreme Court. Most of the pro-environmental decisions of the judiciary uphold the concern that conservation of forest resources and wildlife protection can be made through adherence to laws and quality of environment can be preserved in compliance with law and thereby people can enjoy a healthy environment. Indian judiciary has taken pro- environment standing in some cases. Such as- Bhopal Gas Tragedy 1984 by infusion of pro- environmental steps in the constitution and by declaring right to healthy environment as a fundamental right guaranteed under Article 21 of the constitution. The Court has also directed many companies to take necessary measures to reduce pollution and to protect the environment.

### Pro-Development Approach

Besides, pro- environmental decision, in many cases the Court has derived its own principles for protecting environment, upholding the rule of law. The outcome of Tehri Dam case, Narmada Dam case and construction of Thermal Power Plant at Dahanu Taluk are examples of judicial pro-development approach. The basis for challenge in such cases is adverse environmental impacts, safety issues, rehabilitation measures due to these developments etc. The court's pro-development approach is made to bring development for one section of the society at the cost of another section of the society and environment.

### Polluter Pays Principle

It means the absolute liability for the harm to the environment extends not only to the cost of restoring the environmental degradation, but also to the remediation of the damaged environment, which is a part of 'sustainable development'. The object of this principle is to make the polluter liable to pay compensation to the individual sufferer as well as the cost of restoring the damaged ecology. In Vellore Citizens Case, the principle was adopted to check pollution of underground water caused by tanneries in Tamil Nadu and the polluter pays principle held as part of environmental law of the country. In Vellore Citizens welfare Forum v. Union of India (3SCC), the Supreme Court has directed the High Court to constitute Green Benches to deal with cases related to environment. Now, the green benches are already functioning in some of the high courts like Punjab and Haryana, Calcutta, Madras, Madhya Pradesh and Allahabad High Court.

## Sustainable Development and Interdependence

In Taj Trapezium Case, Supreme Court applied the principle of Sustainable development and to supervise the resultant actions regarding the negative effects that emissions from Mathura oil Refinery had on Taj Mahal. On the direction of Supreme Court, National Environmental Engineering Research Institute (NEERI) and the Ministry of Environment and Forests had undertaken a study for redefining the Taj- Trapezium Zone and re-alienating the area management environmental plan. Supreme Court constituted 'The Mahajan Committee' and directed to inspect the progress of the green belt developed around Taj Mahal.

## Absolute Liability Principle

In the historic case of the oleum gas leak from the Shriram Food and Fertilizer factory in Delhi, in 1986, the Supreme Court ordered the management to pay compensation to the victims of the gas leak. The "absolute liability" of a hazardous chemical manufacturer to give compensation to all those affected by an accident was introduced in this case and it was the first time compensation was paid to victims.

## Public Trust Doctarine

In 1985, activist-advocate M C Mehta filed a writ petition in the Supreme Court to highlight the pollution of the Ganga by industries and municipalities located on its banks. In a historic judgment in 1987, the court ordered the closure of a number of polluting tanneries near Kanpur. Justice E S Venkataramiah, in his judgment, observed: "Just like an industry which cannot pay minimum wages to its workers cannot be allowed to exist, a tannery which cannot set up a primary treatment plant cannot be permitted to continue to be in existence." Similarly, in MI builders Case, a city development authority was asked to dismantle an underground market built under a garden of historical importance.

## Precautionary Principle

Environmental measures must anticipate, prevent and attack the causes of environmental degradation Lack of scientific certainty should not be used as a reason for postponing measures. In Narmada Bachao Case, the Supreme Court held that precautionary principle could not be applied to the decision for building a dam whose gains and losses were predictable and certain.

## Environment Awareness

In M.C. Mehta v. Union of India, to create awareness about the protection of environment, the Supreme Court directed the Central Government to Introduce Environmental Studies as a subject in educational institutions and to observe keep the city clean week. As a part of Environment education, Supreme Court in M.C. Mehta case directed the Union Government was obliged to issue directions to all the State governments and the union territories to enforce through authorities as a condition for license on all cinema halls, to obligatory display free of expense no less than two slides/messages on environment amid each show.

Therefore, there is a need for comprehensive analysis of how law operates as an instrument of environmental protection. The Indian judiciary plays a remarkable role in uplifting the goal of preservation through its various landmark decisions and the Acts and laws which provides a platform so as that one cannot exploit the nature and its gift for his or her greedy needs. The Supreme Court is of view that the objective of all laws on environment should be to create harmony between developments on one hand and preservation of environment on the other hand. The Indian courts have acted on the understanding that loss of natural resources can not be replenished. One who pollutes the environment is liable to pay for it.

With the intervention of judiciary in environmental governance, a number of innovative principles and methods have been introduced to resolve environmental issues and to understand their impact on environmental governance. Few innovative methods are-

- ☆ Right to environment
- ☆ Procedural changes
- ☆ Remedial flexibility
- ☆ Spot visit
- ☆ Evolution of Environmental principles and Doctarines
- ☆ Independent Expert Committee

Thus, Legislature and judiciary plays a catalytic role in uplifting the goal of environmental protection through its various landmark decisions but still there is need to make better laws for preservation. The various laws and acts make available a platform so, that one can not exploit the natural gifts given for his/her greedy needs. No one can imagine his life without nature. Endangering environment endangers existence of human beings. Environment is the gift of nature and we have to respect it. It is the duty of every individual to conserve the beauty of environment.

*"Live Green Love Green Be Green, We are The Mankind Without the Earth We Are Nothing"*

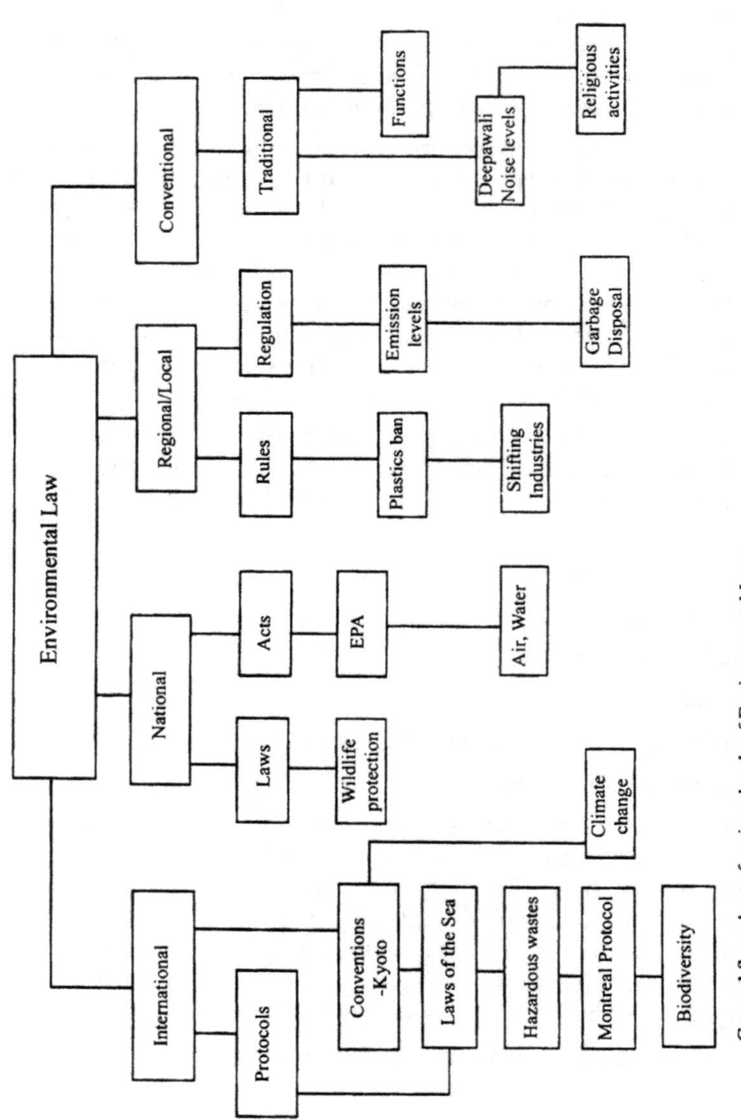


General flow chart of various levels of Environmental Laws.

# Chapter-22

# Marching Towards Sustainability

Earth system is a complex social-environmental system including vast collection of interacting physical, chemical, biological and social components. The biological component of earth system provides environmental process that regulate the functioning of earth. In this system, human beings, animals, plants, water, land and air are closely interconnected and interdependent. We are living in a world in which science, technology and development play important role in changing human destiny. In order to achieve targets of developments, developing and developed countries are irrationally exploiting their natural resources without concerns and understanding of how to exploit and when to exploit. However, over exploitation of natural resources for the purpose of development leads to serious environmental hazards. When life is turned into excessive luxury and comfort, the balance between usage of resources and availability of resources will break. Our life is widely dependent on availability of natural resources. But, we, human beings have no responsibility for Nature. We have destroyed nature and natural resources leading toward environmental damage. Modern world is marked by environmental crisis in which blind race of development is causing stress and strain on environment. Environmental degradation in the world is proceeding at an alarming rate threatening human survival. Among all the problems like rapid population growth, consumption pattern causing stress on natural system , degradation of natural resources affecting life support systems- air, water, soil and biodiversity, global climate change, unemployment, pollution in cities and rivers, polluting industries are we not leading to a disaster ? It is stated that at the current rate of consumption, the earth's resources may soon be reduced to the point at which the living world would be unable to sustain life in the manner that we know. Therefore, natural resources conservation has become one of the most important challenge to face human race.

Environmental sustainability is the most burning issue with which every one of us is concerned. Development should be seen as delicate balance between human needs and nature to fetch positive development and hence emerged the concept of sustainable development. It is most appropriate to begin in terms of Brundtland Report' sustainable development.  It is the development that meets the needs of the present without compromising the ability of future generations to meet their own needs (WCED, 1987). In other words, sustainable development allows a long term utilization of environmental resources for social and economic development while at the same time, attempts to maintain the quality of the environment. Mahatma Gandhi being a man of vision, gave us a new vision to harmonise nature with needs of people. He was in a favour of holistic approach to development for preserving nature and environment. Gandhi without specifically mentioning the word' environment' integrated the whole concept in his thoughts on decentralization, limiting the wants, simplicity and machinery and so on. Environmental consequences of 'Modernization' were well predicted by him and he had aptly anticipated environmental disaster of modern day economic development and this could be traced to his writings in 'Hind Swaraj, 1909'. Advocating the sustainable development, Gandhi once posed the question *"how much a person can consume or how much a person should consume"* when our greed are unlimited but nature's resources have a limit. One can realize the deeper significance of one of his saying *"The country's development has to be in harmony with nature. The earth has resources to meet everybody's needs but not anybody's greed"*. In his opinion, in any scheme of development, man is always at the centre playing important roles  in natural/ environmental system in different capacities. Man has to make a judicious use of natural resources keeping the future generations in mind. The ecological balance should not be disturbed. But, man's contribution has changed from 'user' through 'modifier' and to 'destroyer' of the environment or nature. Man has always tried to cross the limit set by nature and tries to modify the environment according to his need. The most of the environmental hazards we are facing today is the result of non- sustainable developmental policies and strategies. Present day environmental status is the outcome of unlimited exploitation of natural resources. The offshoot of rapid mechanization not only lead to industrialization and industrialization to urbanization to unemployment to poverty, but will also resulted in degradation of environment. According to T.N. Khoshoo, the present ecological breakdown is due to greed of rich people, need of poor people and haywire application of technology. Our lifestyle is based on high level of consumption and it ultimately leads to high environmental damages. Gandhi warned: *"A time is coming when those who are in mad rush today of multiplying their wants, will retrace their steps and say: what we have done?"* Now this warning has become reality.

Sustainable development has multiple objectives. In planning for development, there must be deliberate consideration of how to maintain the quality of the environment, human wellbeing and economic security. The Brundtland comission, in Our Common Future, and subsequently Agenda 21 set out recommendations for developed and developing countries regarding use, housing, waste treatment, transportation and health care. The same advice applies locally. Just as the economic development of a country is linked to its environment and citizens, a community takes into account this interconnectedness in planning for the future.

## Anthropocentric and Ecocentric Approach

According to Soundararaj, the meaning of sustainability sharply varies according to two sets of economists- the Anthropocentrists (who are concerned with economic development for the human) and the Ecocentrists (who are concerned, primarily, with the preservation of the ecosystem). In anthropocentric view, a sustainable state is one in which economic development is sustained through time. It can be done by sustaining utilities and maintaining consumption levels to match them. It also views that it is a state in which resources are so managed as to maintain production opportunities for the future. Similarly, the difference between human centric shallow ecology and bio- centric deep ecology is that shallow ecology takes an anthropocentric( Human centered) view of the world whereas deep ecology is eco-centric giving equal to non human organisms. The concept of sustainability belongs to the social and cultural avenues of life too. In social sphere, sustainability means the establishment of an egalitarian society that ensures equal sharing of resources and opportunities for human. The Rio Declaration states that human beings are at the centre of concerns for sustainable development, and that they have right for healthy life in harmony with nature.

## Sustainable Livelihood Approach

Another approach for sustainable development is creation of sustainable livelihood. It creates goods and services for the community. Consequently, people develop purchasing power with economic and social equity. Therefore, sustainable livelihood is satisfying remuneration that enables every member of the community to nurture the nature and regenerate the resource base.

## Sustainable Community Approach

A sustainable community is one that independently manage its resources and develop a strong social and economic environment for people. A sustainable community should have – living within ecological carrying capacity, protection of natural diversity, clean safe environment, maximum use of people's abilities, minimum use of natural resources, equal opportunities for individual development, recognition of basic responsibilities etc.

## Sustainable Human Development Approach

Sustainable human development approach puts people at the centre of development, protects the life opportunities of future generations as well as the present generation, and respects the natural systems on which all life depends. It is pro- people, pro-jobs, and pro- nature. It gives priority to poverty reduction, 'productive employment, social integration and environmental regeneration'.

## Sustainable Development and Environmental Concerns

Among the ultimate objectives of sustainable development are to achieve a balance between human needs and aspirations in balance with population, resources and the environment and to enhance the quality of life today and in future. A good environment is a fundamental right of every Indian citizen under the right to life

(Article 21). Indian Government has enacted various Acts and laws to meet with various environmental requirements. The Indian Constitution is among the few in the world that contain provisions on environmental protection. Not only in India but for most of the developing countries, too many rules and acts create condition which favour violators. Besides, changing governments and inconsistent bureaucrats keep on amending policies and laws. Ultimately, neither environmental paradigms nor development maintain sustainability. Principle 8 of the Rio Declaration stated that 'to achieve sustainable development and a higher quality of life for all people, States should reduce and eliminate unsustainable patterns of production and consumption and promote appropriate demographic policies'.

A judicious 'sustainable distribution' is the only way to offer all the life- saving basic requirements for all and ensure a socio- economic sustainable environment. Besides, 'sustainable health' is another step for biologically sustainable environment. The increasing rural- urban and urban-urban migrations due to various problems create numerous environmental problems. Therefore, 'sustainable community' should be maintained. Government and its policies should have 'sustainable governance' without which developmental aims are meaningless.

There is urgent need to bring into balance population dynamics, socio-economic development, the use of natural resources and environmental quality. The world civilization must unite to combat the challenges of environmental hazards posed by man. The world must unite to protect the environment. Hence, paying close attention to individual needs, reproductive health, education, equality between woman and man is needed, and a sustainable, equitable and vital balance between human population and available resource is to be maintained.  Let us change our outlook to resist unsustainable practices. The only way  forward is to rethink of conservation vis-à-vis utilization. Use and throw is not fit for sustainable development while REDUCE, REUSE and RECYCLE is the best concept of sustainable development as it refers to balancing of needs and usage. We must care for nature and then only nature will care for us. If we misuse or overuse or abuse the nature, then it will take the very right to life of us. Let us start with ourselves so that future generations do not blame us. Mantra of development should be spirituality with high technology. Both these things allow us to reduce our greed for resources and live in harmony with nature. Gandhi's lesson on *'Protecting and conserving environment along with the developmental processes'* is still relevant in current scenario and age when threat of climate change is looming large and natural resources are fast depleting. Conservation of nature is our moral duty for other living beings and future generations. We shall have to revive the ancient eco-friendly relation to live successfully and satisfactorily according to law. We need a sustainable society which works in partnership with nature and conserve natural resources.

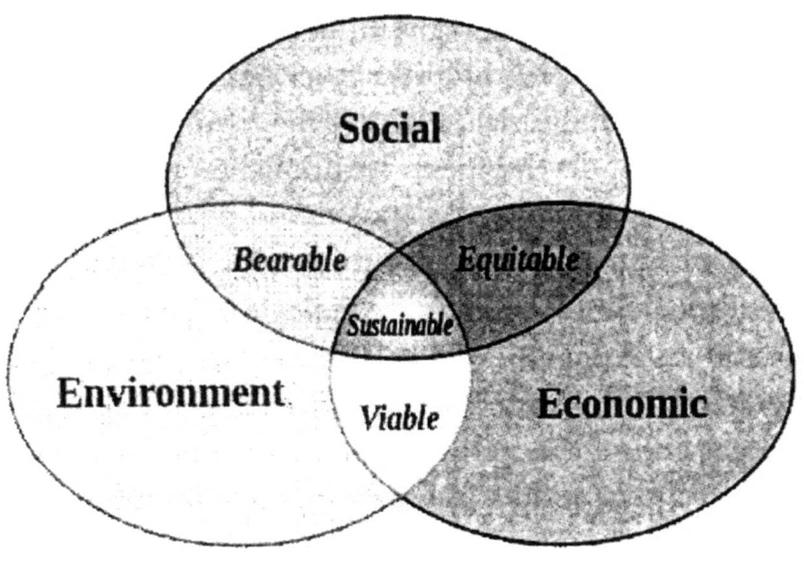



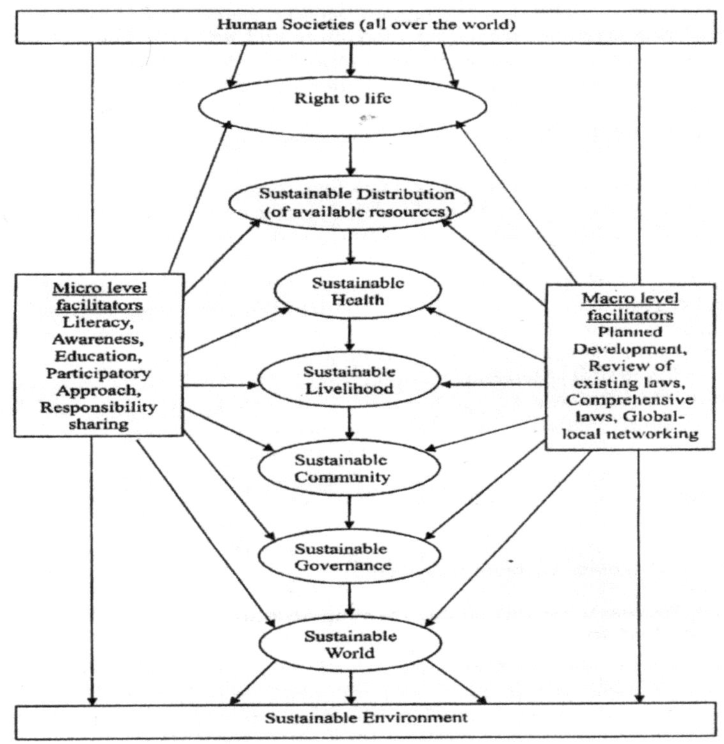



An integrated modal for environment friendly sustainable world

# Chapter-23

# Sustainable Development – New Approaches

The world commission on Environment and Development in its report to the United Nations in 1987 defined the "sustainable development" as "meeting the needs of the present without compromising the ability of future generation to meet their own needs".

Due to uncontrolled population explosion, world wide industrialization, deforestation, urbanization and practices of pollutants control at the "end of the pipe" i.e. at the point where pollutants enter the environment, are no longer effective in controlling the present industrial effluents both in quantity and complexity. Apart from the command control approach of pollution control, pollution is shifted from air to land to water and back again. Therefore, "end of pipe" approach of pollution control is not effective due to increasing quantity and complexity of pollution from 'non-point sources'. With the growing pace of Indian economy, it is very important to ensure the sustainable development. Traditionally there has been a constant and sometimes a bitter struggle between industrial and environmental organizations. Industry prefers to concentrate its technological developments for the benefits of its economic growth, regardless of the pollution it generates whereas environmentalists want the industries to cease and desist pollution at any cost. Thus, there is requirement of new technology for process control, product design and monitoring with simultaneously zero emission and pollution for "sustainability" to occur. Therefore, main emphasis should be on pollution prevention and environmental designing.

## Pollution Prevention

Pollution prevention is a multimedia management to pollution to achieve "front end" reduction of pollutants in the waste by controlling the industrial processes with respect to process upsets, undesirable emissions and by improving the product quality, reducing raw material loss to waste and recycling of byproducts. Pollution prevention eliminates the transfer of pollutants from one media to another using raw material more efficiently. It is also essential to review the overall synthetic sequence in the production of chemical substances and to consider the elimination of hazardous chemicals wherever possible. All the scientific, environmental and economic impacts should be incorporated into the synthetic scheme to ensure environmentally benign synthesis of chemical substance.

## Environmental Designing

Green chemistry program is an initiative under the 'design for environmental program'. It gives equal respect to the cause of both environmental protection and industrial success. Green or sustainable chemistry can help us achieve sustainability in three key areas that could form the basis to protect our environment without harming growth and development. Three areas are- (a) improving the process converting solar energy into chemical and electrical energy; (b) obtaining the chemicals used in the industry from renewable resources instead of obtaining them from oil and petroleum- a fast depleting natural resource; (c) replacing polluting technologies with suitable non-polluting ones.

Green chemistry is the use of chemistry for pollution prevention by environmentally- conscious design of chemical products and processes that reduce or eliminate the use of hazardous toxic substances, solvents and other auxiliaries. It is not only includes the shift of the use of harmful chemicals but also directs to set the usage systems where the use of chemicals could be minimized with maximum economy. Thus, this involves the design and redesign of chemical synthesis with renewable (plant based) rather than non-renewable (fossil carbon derived) raw materials and processes to prevent pollution, rendering the products cost effective as well as environment friendly. Green chemistry provides the best opportunity for manufacturers, processers and users of chemicals to carry out their work in the most economical and environmentally beneficial way. Green chemistry is already demonstrating the potential and promise to develop new techniques and methodologies to enable industries to pursue their traditional innovations and at the same time minimizing environmental impacts. One example is a change in the manufacture of 'Teflon'- non-stick coating commonly used in household cookware. Teflon is traditionally manufactured in water to achieve the necessary chemical reaction. With the help of green chemistry, it was discovered that carbon dioxide works much better to create non-stick coating and leaves little or no waste from the process. The innovations in technology must be planned and implemented in such a way that they are sustainable both economically and environmentally. Now, it is the turn of green chemistry towards sustainable development via pollution control and resource conservation. Basic principles of green chemistry are as-

1.  Environment friendly.
2.  To reduce pollution.
3.  To prevent waste instead of cleaning up.

    It is most appropriate to carryout a synthesis so that formation of waste is minimum or absent. The well known saying *'Prevention is better than cure'* should be followed.

4.  To use and generate less hazardous chemical syntheses with safer outputs.

    Whenever practicable, synthetic methodologies should be designed to use and generate substances that are less or non-toxic to humans and the environment. Heart of green chemistry is the redesigning of existing transformations to incorporate less hazardous materials.

5.  Designing safer chemicals.

    It is now possible because of advancement in understanding of chemical toxicity. A correlation occurs between chemical structure and toxic effects.

6.  Encourages innovation and promotes products that are both environmentally and economically sustainable.

7.  To use safer solvents and auxiliaries.

    Auxiliary substances are used in the manufacture, processing at every step but does not become an integral part of the chemical. The problem of these solvents has been overcome by using such solvents which do not pollute the environment. Such solvents are known as green solvents. For example- liquid carbon dioxide (supercritical $CO$), ionic liquid water.

8.  To design energy efficient chemical processes.

    Energy requirements should be minimized for environmental and economic impacts. It can be kept to a base minimum in certain cases by use of catalyst.

9.  To use renewable rather than depleting Feedstock.

    Non-renewable or depleting sources can exhausted by their continual use. So, these are not regarded as sustainable from environmental point of view. The starting materials obtained from biological or agricultural processes are renewable starting materials. Carbon dioxide obtained from natural sources and methane gas from natural sources like marsh/ natural gas are available in reasonable amounts and therefore, considered as renewable starting materials.

10. To use superior catalytic reagents as selective as possible.

    Another Green Chemistry approach is the use of a catalyst which facilitates transformations without the catalyst being consumed in the reaction and without being incorporated in the final product.

11. To reduce unnecessary derivatives.

    A commonly used technique in organic synthesis is use of protecting or  blocking group. The groups used may make the reaction to go in an unwanted way, if it is left unprotected. This procedure adds to the problem of waste disposal.

12. To design for degradation products not persisting in the environment.

    The products should be biodegradable and should not be persistent chemicals. It is also important that degradation products do not possess any toxicity and detrimental effects to the environment. Plastic, pesticides (organic halogen based) are examples which threaten the environment.

13.  Inherently safer chemistry for accident prevention.

Accident in Bhopal (India) and Seveso (Italy) and many others are the examples which resulted in thousands of life loss. It has been found that in an attempt to recycle solvents from a process increases the potential for a chemical accident or fire (may be for economic reasons). Substances used in chemical process should be chosen so as to minimize the potential for chemical accidents, explosions or fires etc.

14.  To conserve raw materials.

15.  'Green' processes are more cost -effective than conventional methods.

16.  Using production methods with fewer steps.

## Why Green Chemistry?

Green Chemistry is effective in reducing the impact of chemicals on human health and the environment. In addition, many companies have found that it can be more cost effective and even profitable to meet environmental goals. Its higher efficiency, less waste, better products, reduced unnecessary derivatives results in profits. It allows companies to comply with the law in much simpler and cheaper manner. Addressing the problem of hazard at the molecular level, it may be applied to all types of environmental issues. Green organic chemistry has led to more environmentally sound agricultural processes. Safer pesticides are also being developed. For example- Spinosad, an insecticide manufactured by fermenting a naturally occurring soil organism, was registered by EPA as a reduced-risk insecticide in 1997. Spinosad does not leach, bioaccumulate, volatize or persist in the environment and in field tests left 70-80% of beneficial insects unharmed. It is less toxic to mammals and birds while moderate toxic to aquatic organisms. Another example may be cited for Ethyl lactate, an industrial cleaning solvent made from cornstarch and soybean oil patented in 2000and is competitively priced with petrochemical solvents. It biodegrades to carbon dioxide and water having no harmful impact on the environment, human beings and also wildlife. Similarly, many industries, such as the pulp and paper industry, use chlorine compounds in processes that generate toxic chlorinated organic waste. A new technology has developed through green chemistry that converts wood pulp into paper using oxygen, water and polyoxometalate salts, producing only water and carbon dioxide as by-products. EPA supports the development and use of innovative chemical technologies to manufacture consumer products that prevent or reduce pollution in academia and industry.

## Emerging Concepts

Several new concepts and industrial practices are emerging to facilitate the integration of industrial production with environmental factors.

☆  ***Concurrent Engineering:*** Environmental factors can be considered as an element of the concurrent engineering "Design for X" process. 'X' represents a product characteristics such as reliability that the company wants to maximize in its product design.

- ✮ **Total Quality Management (TQM):** The adoption of TQM programmes has rendered the quality as the responsibility of everyone in the company from top to bottom.

- ✮ **Life Cycle Approach:** The environmental impacts including safety, health and social factors, should be considered across the life time of the products, process, material, technology and services. This approach is meant for minimizing the environmental impact of a product.

- ✮ **Preventive Environmental Management (PEM):** It involves elimination of wastes and pollutants at their sources rather than at the end-of –pipe stage. Since the pollution occurs due to anthropogenic activities, PEM poses a challenge to all sections of society, at all levels of activity and decision making. Thus, PEM requires environmental awareness.

- ✮ **Regulation on Occupational Safety and Health (OSH):** India and all other countries should promote occupational Health and Safety by enacting laws and acts to regulate the measures taken by the companies for health and safety of their employees.

- ✮ **Energy Utilization:** Energy is required to support multifarious activities of the civilized society. However, the actual amount of energy required depends on the efficiency with which it is used, which in turn depends on technology, the integration of energy systems, energy conservation practices, and on our living pattern. When energy is converted from one form to another, there is always loss of some useful energy as waste heat. Minimizing the waste heat and maximizing the utilization represents a major challenge in energy conversion.

Now, it is the turn of green chemistry towards sustainable development via pollution prevention and resource conservation. It not only helps us in designing of new ways to synthesize the desired product economically, user friendly and is also environment friendly. Green chemistry is also one such step taken towards sustainability and well being of human race. Government should also make some strict rules in governing the industries to use eco-friendly ways of production. There is a need to maintain balance between supporting new developments in chemistry and bringing the previously established chemical processes under greener and sustainable way.

# Chapter-24

# Changing Environment: Adaptive Strategies and Technologies

Environment, on which all human beings depended for their survival, is variable, dynamic and changing because of the forces within itself and also the activities of human beings. The problem of today, such as drought, forest fires, flooding are magnified by climate changes. This climate change is resulting in extreme weather events that are disproportionately more severe in developing countries that do not possess the capacity to adapt quality to such variability. From the time immemorial the growth and activities of human habitations, they have had their impact on natural resources as much as the natural resources have impacted the human societies. The changing environment mainly refers to the changes in precipitation pattern, temperature, soil stock, water quality, siltation, forest cover etc. These changes have impacted the occupational and cultural patterns among communities which have been traditionally depending on the resources of the existing natural environment. Therefore, the traditional occupations and cultural patterns are directly affected by changing environment.

## Adapting to Changing Environment: Why and How?

In many parts of the globe, however, future impacts of climate changes are likely to be significantly more than those that have been experienced in the past as a result of natural climatic variability alone. Therefore, communities have started reshaping their traditional life with occupational, environmental and cultural changes. As being directly affected, various communities have opted using various strategies and technologies to adapt themselves with the changing environment at different times and in different places. These adaptive measures may be undertaken by individuals, communities, private sectors, civil societies and government and will consists of a wide range of behavioral, institutional , structural and technological adjustment.

These adaptive strategies can be – 'Planned' and 'Autonomous' types.

'Planned' adaptation is the result of decisions that are based on an awareness that the   conditions have changed - or are about to change - and that some type of action is mandatory to achieve, maintain or return to a desired and required state. For example- Building sea walls in anticipation of a rise in sea level.

'Autonomous' adaptation refers to changes that natural and most human systems undergo in response to the changing conditions in their immediate environment, irrespective of any broader policy or plan based decisions.  For example- Shifting market signals, price of crops, changes in farming practices, changes in recreational and tourist's behaviour etc.

In many cases, people will adapt to climate change simply by changing their behavior, by moving to a different location or by changing their occupation. But, sometimes they employ different forms of technology, whether 'hard' forms such as new irrigation systems or drought resistant seeds or 'soft' technology such as insurance schemes or crop rotation patterns, or they may use a combination of hard and soft technologies both. It should be possible to adapt to some extent by modifying or extending existing technologies. Adaptation on the other hand involves coping with changing environment- taking measures to reduce the negative effects or exploit the positive ones, by making appropriate adjustments.

## Adaptive Strategies include

### 1- Water Adaptation

- ☆ To improve water resource management including flood risk and drought control.
- ☆ To raise awareness by engaging with water utilities and water users.
- ☆ Water monitoring programmes.
- ☆ To conserve water.
- ☆ To adopt water harvesting, increased water storage and reduced leakage.
- ☆ Watershed management.

### 2. Agriculture Adaption

The adaptive strategy needs and measures for agriculture differ clearly in developed and developing countries. More focus should be on adapting agriculture in developing economies rather than in developed ones.

- ☆ To develop more efficient irrigation techniques, new cultivars, changed cropping patterns etc.
- ☆ Mangrove and reef protection measures.
- ☆ To educate public.

### 3. Fisheries Adaptation

- ☆ To develop marine protected areas.
- ☆ To develop alternatives- for example, inundated coasts to aquaculture.

## 4. Tourism Adaptation

☆ To incorporate climate change concerns in the development of tourism plan.

☆ To promote energy efficiency, water conservation and use of renewable energy.

## 5. Coastal Zone Adaptation

Coastal zone risk increased damage from flood and storms and if global average temperature goes up by $3^0C$, global coastal wetlands will be lost and millions of people have to face coastal flooding every year.

☆ To change aquaculture practices.

☆ To improve regulations for restricting coastal developments.

## 6. Ecosystem and Forest Adaptation

The adaptation of natural ecosystem is very closely related to other strategies such as mangrove conservation and forest management. Efforts should be focused on-

☆ To promote awareness in society.

☆ To promote community based conservation programme.

Nowadays, human society can also take advantage of 'high' technology such as earth observation systems that can provide more accurate weather forecasts or crop that are based on genetically modified organisms. But, adaptive measures should be more flexible and adaptable to local circumstances which mean that in addition to being socially and legally acceptable, they should be reasonably cost effective.

## Some Examples of Adapted Strategies

Technology can be a panacea for dealing with impacts of changing climate on coastal zones. The technology adaptation may form a vital part of frameworks of integrated coastal managements. In Bangladesh, every year, millions of people have to face flood. Therefore, Bangladesh has been developing an effective early warning system to adapt to risk of flood and cyclone through a system of flood monitoring and forecasting. Among these is a five year project, the Community Flood Information System (CFIS), which is funded by United States Agency For International Development(USAID). CFIS is designed to help communities to adapt to future risk and to protect crops, habitats and livestock.

Rise in sea water temperature are continuously damaging the coral reefs. The local communities have started adapting new technology for restoring reefs by making 'Bio-rock'. It is created by a low voltage electric current through sea water causing dissolved minerals to precipitate on the surface that finally grow into white limestone structure; similar to the material that make up coral reefs and nourish tropical white sand beaches. These 'Bio-rocks' has been found to accelerate coral growth in areas facing environmental stresses. This also help in forming structures that have populated by a range of coral reef organisms such as fish, crabs, octopuses, lobsters, and sea urchins etc.

Agricultural producers have also modifying their practices to cope with climate change stresses and variability. Farmers have taken advantage of latest technologies to face more easily with arid environment introducing new crop hybrids and making use of scarce water, as with system of drip irrigation. Much increase in yields in recent decades resulted from improved management practices, irrigation patterns, increased use of fertilizers or other inputs but around half was due to genetic improvement in crop variety through biotechnology. Introduction of transgenic varieties of seeds are on the forefront of usage in spite of their higher cost because of having specially acquired traits like herbicide resistance. This trait of herbicide resistance has a chance to give higher yield without spraying herbicides. Besides, introduction of elite clone derived plantlets or callus derived saplings, in the field of floriculture, horticulture, medicinal plants and even in forestry, has restored high degree of yield and productivity. Mukund  Research Foundaton, Coimbatore has achieved remarkable success in producing dwarf , multipurpose and convenient to package Papaya. The same master stroke is possible in many others such as oranges, grapes, apples, banana etc. Thus tissue culture technique has brought enhanced yield of comparable quality of many agricultural, horticultural and floricultural produce.

Farmers, living in areas which remain inundated for long periods in Bangladesh, have adopted 'Floating Agriculture'- a system  similar to hydroponics, in which plants derive their nutrients from water instead of soil. The floating beds are mostly used to grow vegetables are more fertile and more productive. To make beds of this 'Bio-Land' farmers use water hyacinths, aquatic algae, water warts or other water borne creepers along with straws and herbs or plant residues.

The changing environmental conditions of the Shonebeel fishery of Karimganj district of Assam have played a major role in the emergence of new strategies the fishing communities are adopting. The climate change, deforestation, siltation, pollutants and garbage disposal are adversely affecting the land, forest, water and fish resources in the beel. The villagers of Shonebeel are adopting migration, occupational diversification, new agricultural and fishing practices, alternative fuel resources and cultural accommodation as the strategies for their survival.

Nowadays, human society can also take advantage of 'high' technology such as earth observation systems that can provide more accurate weather forecasts or crop that are based on genetically modified organisms. But, adaptive measures should be more flexible and adaptable to local circumstances which mean that in addition to being socially and legally acceptable, they should be reasonably cost effective. Now, there is need to implement policies for planned adaptation aimed at the impact of changing environment and at facilitating, preparing, complementing autonomous adaptation initiatives. Modern technologies and strategies, therefore, would be based on definite relationship between cause and effect, quantifiable in terms of short term as well as long term advantages and disadvantages. There is a dire need for modification in our age old practices in such a way that every input advances our step towards sustainability in current changing environment. In 21ˢᵗ century, we will have to shed our old age practices and give more weightage to bio-tech inputs which are cost effective, eco-friendly and sustainable.

# Bibliography


Agarwal Bina. (1993). The Gender and the Environment Debate: Lessons from India. *Feminist Studies*. Vol. 18, No.1.,

Ambwani Sunil. (2007). Environmental Justice: Scope and Access, A*IR, Journal 49, Operation and maintenance of the plant.*

Asif Ahmad. (2011, Feb.). Indian Environmental Concerns: Protection of Indian Religious and Cultural Heritage. Peace and Collaborative Development network.

Bast Felix (2018). Paradise lost: How climate change and pollution could wreak Havoc in Antarctica ? *Science Reporter*, June, pp. 36-39.

Behari, J. (2010). Biological responses of mobile phone frequency exposure. *Indian Journal of Experimental Biology*, 48: 959-981.

Bell Daniel. (1973, Peris). Technology, Nature and Society, American Scholar, Summer.

Bhardwaj Rakhi; Sharma Rahul . 2011. Effect of Electromagnetic Radiation On man and its surroundings. ABS/A/07. pp 4, section A in 1st World congress for man and nature "Global climate change and Biodiversity conservation" Invention of Research , 11-13 Nov.2011. Gurukul Kangri Vishvidyalaya, Hardwar, Uttrakhand.

Bilani, H.U. (1977). Urban Problems. IIPA, New Delhi.

Bose, A. (1978). India's Urbanisation. Tata McGraw-Hill Publishing Co.Ltd. New Delhi.

Braatz, S.M. (1997). State of World Forests. *Nature and Resources*. UNESCO, 33: 24-65.

Chandrappa, Ramesh and D.R. Ravi. (2009). Environmental Issues, Law and Technology- An Indian Perspective. Delhi Research India Publication.

Chawla R. (2005).Tourism and Biodiversity.  Sonali Pub., New Delhi.

Chemical cuisine, *Consumer voice*, Jan.2009, pp 26-27.

Cohen Joel E. (1995), How many people can the Earth support. New York. W. W. Nortan.

Davies Kingsley. (1962). "Urbanisation in India- Past and Future", in Turner, R. (*ed.*) India's Urban Future, University of California Press, Berkley.

Dewan, M.L.(1987). Water Harvesting. *Indian J. Soil Cons.*, 15(1): 30-34.

Divan,S. and Rosencranz, A. (2001). Environmental Law and Policy in India, Cases, Materials and statutes, New Delhi, India.

Dugger, Celia (2006). "Need for water could double in 50 years, U.N.Study finds". New York Times, August 22.

Faisal, A.M. (2008). Climate change and phonology. New Age, 3 March.

Gandhi, M.K. (1938). Hind Swaraj. Navjivan Publishing House, Ahemdabad.

Gokak, V.K.(1979). A Value Orientation to System of Education. New Delhi: M.M. Gulab and Sons.

Govt. of Assam (2006). Record of the Regional Agricultural Research Station, Karimganj  District, Karimganj, Assam.

Gurjar, R.K. and B.C.Jat. (2002). Man and Environment. Jaipur: Panchsheel Prakashan.

Herper,C. (2004). Environment and Society: Human Perspective on Environmental Issues, New Jersey:Pearson Education.

Houghton, J.T.; Jenkins, G.J.; Ephraums, J.J. (Ed.). (1990). Climate Change: The IPCC Scientific Assessment, Cambridge University Press, Cambridge.

IPCC (Intergovernmental Panel on Climate Change), 2007, Summary for Policymakers In: climate change 2007: The Physical Science basis; Cambridge, United Kingdom, Cambridge University Press.

Jaidev, Somesh, (2010). Natural Resources in 21ˢᵗ Century. Oxford Book Company, Jaipur.

Jain, A.K. (2013). Eco- Technology and Bio-Architecture for sustainable development. *Envis Journal* (SPA), New Delhi.

Jat,M.L.; O.P.Gill; B.S.Deora and Vivek Kumar. (2017). Rainwater Management Theory and Practice. Scientific Publishers, Jodhpur.

Jha, Birendra Kumar and Pradeep Kumar Tripathi,(2014). Food Security in India: A Challenging Issue. *Bihar Economic Jounal*, vol. 3, No. 1. August. ISSN 22308970.

Kiran Rekha.(2007). 'Value and Environment', Education in Human Value, New Delhi: Knashka Publishers.

Khoshoo,T.N. and J.S. Moolarkattu. (2010). Mahatma Gandhi and the Environment: Analysing Gandhian  Environmental Thought. New Delhi, TERI

Lahiri  A.S. (2005). Tourism Industry and Its Impact on Global Environment Independent ,Oct. 1.

Leelakrishnan, P. (2008). Environmental law in India, Mumbai, Lexis Nexis, Third Edition.

Manisha, Tyagi. (2011).Environmental Wisdom in Ancient India. Variorum *Multidisciplinary e- research journal*, vol. ii, issue iii, February.

McNeely, J.A.(1990). 'How Conservation Strategies Contribute to Sustainable Development'. *Environmental Conservation*, vol.17, 9-13.

Misra, Sib Ranjan. (2010). Climate Change and Food security: Some Issues and Challenges In The Proceedings of the International Conference organized by the department of Environmental Sciences, Visva Bharati, Santiniketan, pp. 153-168.

Mooney,H.A.; Winner,W.E. and Pell, E.J. (1991). Response of plants to multiple stresses. Academic Press, San Diegn, California, USA.

Murty, V.V.N.(1982). Land and Water Management Engineering. Kalyani Publication, New Delhi.

Ozha, D.D. and V.N. Mathur. (2007). Revival of Swadeshi Vedic and Traditional Wisdom Of Water Management. An effective tool to mitigate water crisis. In National Seminar on 'Water and Culture'. Hampi, Bellary District.

Patil, S.Y. (1998). Environmental Impact on Urbanisation in India, in Essays on Regional Science, Regional Development and Planning, Vol. II, RSAI, Calcutta.

Ram Prakash.(2003). Environment: Global Changes and Challenges. ABD Publishers, Jaipur (India).

Ranjan Malvika. (2010). Living in Harmony with Nature' Message of Ancient Hindu Spiritual texts. *Educational Quest*, vol. 1; issue1.

Rank, J.(2012). The Effects of Radiation Pollution on the Environment. http/www. Ehow.com/about_5552531_effects-radiation-pollution-environment.html.

Reddy, N.K. (1979). Man, Education and Values, Delhi: B.R. Publishing Corporation.

Rowland, F.S. and Isaksen, I.S.A. (1995). The Changing Atmosphere. John Wiley and Sons, Chichester, pp. 281.

Roy, Shruti (2007). Impact of Tourism on environment. wwwthehindu.com/mag.

Sahu, G.A.(2007). Environmental Governance and Role of Judiciary in India', Bangalore Institute for Social and Economic Change, 24.

Sandhu, R.S. (2003). Urbanization in India. Sage Publication.

Sandler, Ronald. (2007). Character and Environment: A Virtue Oriented Approach to Environmental Ethics, New York; Columbia University Press.

Sen, A.K. and Jean, Dreze.(1989a). Hunger and Public Action, Oxford University Press, New Delhi, pp, 35-42.

Sharma, S.K. (2001). 'Governance for Realising a Sustainable Society', *Social Change*, 31(1&2):165-173.

Shastri, C.S. (2005). Environment Law, 2nd ed. New Delhi: Eastern Book Company.

Shiva Vandana. (1998). Staying Alive: Women, Ecology and Survival in India. London Zed Books.

Shrivastava, D.P. (2010). The role of Judiciary in environmental protection; Bilaspur; High Court of Chattisgarh.

Shinde, P.G. Water Scenario, 2025. In Nat. level Conference on ' water Management Scenario 2025- Problems, issues and Challenges'. pp.1-3.

Singh, P.K. (2000). "Watershed Management( Design and Practices)". 1ˢᵗ Edition. E-media Publications, Udaipur, India.

Singh Savita. (1999). Global Concern with Environmental Crisis and Gandhi's Vision. New Delhi. A.P.H. Publishing Corporation.

Shobha Rastogi; Latika Jha. (2011). Chemistry for Green Environment. Oxford Book Company, Vishveshria Nagar, Gopalpura Road, Jaipur.

Somesh Jaidev. (2010). Natural Resources in 21ˢᵗ century. Oxford Book Company, Vishveshria Nagar, Gopalpura Road, Jaipur.

Soundararaj, F.( 2004). 'Education, Sustainable Development and Educational Management', In Quality Higher Education and Sustainable Development. NAAC Decinnial Lectures.

Southwick, C.H., (1996). Chapter 15: Human Populations "Global Ecology in Human Perspective, Oxford Univ. Press, 159-182.

Subramanian, V.A. (2002). Text Book of Environmental Science, Narosa Publishing House, New Delhi, Chennai, Mumbai, Calcutta.

Tripathi,G. (2002). Man, Environment and Peace; in Bioresources and Resources. Ed. By G. Tripathi and Y.C. Tripathi

UNO, (1995). 'Consumption Pattern World-Wide Report'. New York: United Nations Organisation.

Vasudev, S.P. (2010). 'Integrated and Sustainable Management of Natural Resources'. *Kurukshetra*, March, vol. 58, No.5, pp.8-12.

Vedic way to beat pollution. (1985). The Hindu- English Daily, 4 May.

Wagner, Luca. N. (2008). Urbanisation: 21ˢᵗ century Issues and Challenges. Nova Publisher.

Weaver, J.H.; M.T. Rock and K. Kausterer. (2003). Achieving Broad-Based Sustainable Development: Governance, Environment and Growth with Equity, Jaipur. Rawat Publications (Reprint in India).

Wheeler, W.M. (1916). Ants carried in a floating log from Brazilian mainland to San Sebastian Island. *Psyche*, 23; 180-183.

Whelan, T. (1991). Nature Tourism. Managing for the Environment. Island Press, Washigton, DC ed.

Wolverton Bill, NASA Study; House plants clean air. Zone 10.com.

World Wide Fund (WWF). (1992). India's Wetlands Mangroves and Coral Reefs. Prepared by WWF For Nature India, New Delhi.

www Avianweb.com/chemicals in food.

www sustainable table.org/385/additives.

www natural health365, com. Pesticides on food causes life threating allergic reactions.

Zahn Marcus. (1995). Electric and Magnetic fields in the environment. Wiley Encyclopedia Series in Environmental Science, Energy, Technology and the Environment, vol. ii; Editors-Attilio Bisio and Sharon Books, John Wiley & Sons, Inc. New York.

# Index

**R**

Radiation balance  41

Radiation effect  48

Radiation energy  49

Radiation Exposure  49

Radiation Pollution  47

Rainforests  26, 43

Rainwater Harvesting  95, 99

Rainwater Management  95

Renewable Resources  81

Ripening Agents  70

River Ganga  19

Rock cycle  2

Rooftop rainwater harvesting  99

**S**

Salinization  82

*Sansevieria trifaciata*  67

Sick Building Syndrome  66

Silent killers  63

Smart cities  14

Smart Sustainable Cities  15

Soil erosion  6

*Spathiphyllum wallisii*  67

Specific Absorption Ratio (SAR)  49

Spice Tourism Village  30

Succession  83

Supercritical CO  131

Suspended particulate matter (SPM)  12

Sustainable forest management  7

Sustainable land management  31

Sustainable Livelihood  125

Sustainable Tourism  30

Symbionts  13

**T**

Technology  90

Teflon  130

Thermal damage  49

Total Quality Management  133

Totemism  102

Tourism  26, 89

Tourism Adaptation  137

Tourism Industry  25

Toxic Chemicals  67

Trampling  28

Transport  89

Typhoid  8

**U**

Undernutrition  78

Urban forests  12

Urbanization  11

Urbanization Stress  11

Urban migration  11

**V**

Visual pollution  27

**W**

Water  83

Water Adaptation  136

Water Harvesting  97

Water logging  84

Water occupies  91

Women  107

World Food Summit  75

World Heritage Sites  25

World Travel And Tourism Corporation (WTTC)  31

WWF  29

**X**

X-rays  47